N(

The Nuclear Lion

What Every Citizen
Should Know About
Nuclear Power and Nuclear War

The
Nuclear Lion

What Every Citizen
Should Know About
Nuclear Power and Nuclear War

John Jagger

Plenum Press • New York and London

Library of Congress Cataloging-in-Publication Data

Jagger, John.
    The nuclear lion : what every citizen should know about nuclear
power and nuclear war / John Jagger.
        p.   cm.
    Includes bibliographical referencs and index.
    ISBN 0-306-43771-6
    1. Nuclear energy--Popular works. 2. Nuclear warfare--Popular
works.   I. Title.
TK9146.J28   1991
363.17'99--dc20   91-2095
                                                               CIP

ISBN 0-306-43771-6

© 1991 John Jagger
Plenum Press is a division of Plenum Publishing Corporation
233 Spring Street, New York, N.Y. 10013

Printed in the United States of America

This book is dedicated to
Thomas, Yvonne, Nancy, and Alexander
. . . and to all children, everywhere.

Preface

> *. . . human kind cannot bear very much reality.*
>
> T. S. Eliot, *Four Quartets*

When I was a little child, I lived in an old and somewhat rickety house by the sea. When the winter wind blew, the house would shake and tremble, and cold drafts would whistle through cracks in the walls. You might have thought that lying in bed in a dark room on such cold, windy nights would have frightened me. But it had just the opposite effect: having known this environment since birth, I actually found the shaking of the house, the whistling of the wind, and the crashing of the sea to be comforting, and I was lulled to sleep by these familiar sounds. They signaled to me that all was right with the world and that the forces of nature were operating in the normal way.

But I did have a problem. On the dimly lit landing of the staircase leading up to my bedroom, there was a large and dark picture of a male lion, sitting as such lions do with his massive paws in front of him and his head erect, turned slightly to the right, and staring straight out at you with yellow blazing eyes. I had great difficulty getting past that lion. Someone would have to hold my hand and take me up to bed, past the dreaded picture. Later, the lion entered my dreams, and I had nightmares in which I would casually look out the window at night and be startled by the sight of a lion roaming the beach down where the waves broke on the sand. I would awaken in a cold sweat.

The things that I lived with every day and "understood" or "felt in my bones" did not alarm me. The crashing of the surf is a restful sound to those who live by the sea. It has nothing to do with potential dangers, for even as a child I respected the power of the sea. I knew how gale-whipped waves could break up a stout seawall or smash a boat to smithereens. But I did not generally fear the sea. Not fearing it, I could appreciate its beauty and admire its strength. Like men who gain their livelihood from the sea, I understood it and I respected it.

The lion was a different story. There were no real lions in my world. And the pictures I had seen of them were absolutely frightening. I feared the unknown, especially when it seemed dangerous. Lions were for me a symbol of unbridled, mysterious, and fearsome power.

This book is about the nucleus of the atom. The very fact that we know the atom *has* a nucleus, and that we have discovered how the nucleus can break down, either relatively gently as in radioactivity, or with immense power as in nuclear fission, is almost incredible. Modern physics is a beautiful subject, and the insight it has given us into the nature of the universe is a great achievement of the human intellect. Today, when we look back to the Middle Ages, political figures do not stand out as much as philosophers like Roger Bacon and scientists like Copernicus, because of the tremendous impact they had upon the ages that followed—an impact far more lasting and pervasive than that of mere kings and princes. People in the distant future will forget Bush and Gorbachev, but they will remember Einstein. They will most certainly remember when we first released nuclear energy. As members of the human race, we should be proud of our scientific achievements, just as we are proud of the artistic achievements of Shakespeare and Mozart.

The stupendous energy in the atomic nucleus can be used to advance human welfare, and it has been so used ever since we learned how to release it. Nuclear medicine has revolutionized medical diagnosis and treatment, notably in dealing with cancer. Nuclear reactors have provided us with valuable radioactive atoms (radioisotopes) for use in research and industry, and they have given us cheap, clean power, which can drive a ship around the world on a tiny charge of fuel.

On the other hand, we have unleashed the awesome power of nuclear weapons, and we must now face the almost incomprehensible devastation that awaits the world as it contemplates nuclear war. An all-out nuclear war would end modern civilization, and might well end humankind, to say nothing of countless other species of plants and animals. It would be, without question, the greatest disaster of the last million years of the history of the Earth.

For me, nuclear energy is much like the sea. I can recognize its beauty and its potential for good. At the same time, I see its destructive power and its potential for evil. I feel that I can look at both of these things in a reasonably objective way and make rational decisions about them. This is true of most people who have been trained in the natural sciences.

But to the general public, nuclear energy is a lion. It is a fearsome and terrible thing. And, like my fear of the lion as a child, this fear has arisen because it is not understood. This fact should not be surprising. After all, nuclear physics is not an easy subject to grasp. Furthermore, the engineering problems of nuclear waste disposal, or the political problems of nuclear arms control, cannot be understood without some study.

Yet there is a singular resistance to learning about the nuclear world that goes beyond the mere difficulty of the subject matter. People fear nuclear war to such an extent that they are often unwilling even to contemplate it. This fear can be carried over to all things nuclear, including nuclear power, nuclear waste disposal, and even the very minor problems associated with medical nuclear wastes. Such irrational fear must be overcome. Although we still face daunting problems, many people seem totally unaware of the "good news": international arms agreements have been remarkably successful, such agreements can be readily verified, and nuclear waste can be disposed of with great safety.

This book is a plea for nuclear sanity. The release of nuclear energy has presented humankind with a terrible reality, a challenge greater than it has ever had. We can no longer ignore this challenge. We shall not survive unless we develop mature attitudes about nuclear energy. But we must think about both the good and the evil of nuclear power and nuclear weapons and try to arrive at sound political judgments about them. In short, we must educate ourselves about the realities of nuclear energy.

This book attempts to explain what the nuclear lion is all about. I shall try to lead you up the stairs, right past the lion, so that you can see him up close. I am not going to tell you that he is not fierce, because he is. But I would remind you of a lady who lived in Africa with a lion called Elsa, of whom she made a dear friend, and of how that understanding of the lion enriched her life and made her see and respect a face of nature that she had not recognized before. Lions can be ferocious and they can kill you, swiftly and terribly. But they generally do this only if you disturb or mistreat them; normally, they would much prefer not to harm humans at all. Those who understand this may begin to see the beauty and power of the lion, and to realize that perhaps we shall lose something of great value if we kill the lion just because it can be dangerous.

In short, I shall try to persuade you that nuclear energy can be a force for immense good as well as for immense evil.

Acknowledgments

I am indebted to Vivian Castleberry, journalist and president of Peacemakers, Inc., in Dallas; Ernest C. Pollard, emeritus professor of biophysics, Department of Molecular and Cell Biology, Pennsylvania State University; and Brandy Walker, M.D., nuclear medicine specialist, University of Texas Southwestern Medical Center at Dallas, for reading and criticizing the entire manuscript for this book. Large segments of the manuscript were also read and criticized by Eugene J. Carroll, Jr., Rear Admiral, USN (ret.), deputy director, Center for Defense Information; my wife, Mary Esther Gaulden, radiation biologist, University of Texas Southwestern Medical Center at Dallas; the late Earlene A. Rupert, geneticist, Clemson University; Alvin M. Weinberg, nuclear physicist and originator of the pressurized-water reactor, Oak Ridge Associated Universities; and Virginia P. White, science administrator and free-lance writer from New York City. These experts have been of tremendous help, but of course any errors or misinterpretations that may occur in the text are entirely my own responsibility.

The late Carolyn L. Galerstein, Dean of the School of General Studies, University of Texas at Dallas, was always supportive and encouraging. I thank Bob Ubell, of Robert Ubell Associates in New York, for suggesting the title *The Nuclear Lion*, and Ruth Ricamore for the line drawings. The excellent editing by Naomi Brier, of Plenum Publishing Corporation, was invaluable.

Finally, I am deeply indebted to my wife, both for her professional input and for her patience and encouragement throughout the writing of this book.

Contents

Introduction: Chernobyl and Hiroshima

In the early hours of the morning of 26 April 1986, a nuclear power reactor in the Ukrainian town of Chernobyl blew up, killing 31 people and strewing radioactive debris across Russia and Eastern Europe. The whole world was deeply frightened by this accident. It is now widely felt that Chernobyl sounded the death knell of commercial nuclear power.

On that fateful morning, the reactor operators were conducting an experiment designed, ironically, to improve plant safety. They made six errors that the Soviets characterized as "unbelievable," causing the reactor to go out of control. The reactor core became extremely hot and ignited the graphite moderator (the material that slows down the neutrons). This resulted in a steam explosion and a hydrogen explosion that blew the roof off the building, projecting radioactive materials high into the atmosphere. For the next 10 days, the very hot reactor continued to release radioactive material.

At the power plant, 237 workers and firefighters were exposed to fire and radiation, and many of them experienced acute radiation sickness. Two people were killed almost immediately; 29 more people died in the following weeks. But most of the survivors are now back at work.

A cloud containing radioactive particles and gases from the explosion drifted toward the northwest. In the first 20 miles, it dropped most of its heavier radioactive particles, containing the major portion of the radioactivity. This exposed Soviet citizens in the Chernobyl area to levels of radiation ranging from quite low to moderately high. All of these people, some 115,000, were evacuated and are only slowly returning to their homes.

The remaining radioactive cloud was much less dangerous and was widely dispersed, spreading within the next few days to the northwest as far as Sweden and Britain, and to the southwest as far as Italy. Although the doses received by people in these countries were quite low, very large populations were involved. As a consequence, it is estimated that, in Europe and the European part of the Soviet Union, 17,000 people will die prematurely of cancer over the next 50 years, partly from direct exposure to radiation and partly from ingestion of contaminated food and water.

Ten days after the Chernobyl accident, a World Health Organization committee of European experts stated that there was "no reason for travel restrictions between countries, with the obvious exception of travel to the immediate surroundings of the accident site." In spite of such assurances, people were wary of travel to Europe for months afterward. They did not believe the experts.

The Chernobyl accident aroused tremendous public apprehension about nuclear power. It raised a multitude of questions, including:

1. How great a disaster was it? The death toll was only 31, small in comparison with that in other disasters. But the prospect of 17,000 cancers over the next 50 years is daunting. Is this too great a price to pay for nuclear power?
2. Is commercial nuclear power safe or dangerous? After all, this is the only civilian nuclear-power-reactor accident that has ever killed anyone, anywhere in the world.
3. The Chernobyl accident was in Russia. But we had the

Three Mile Island accident in Pennsylvania in 1979.
Doesn't that show that we are vulnerable even in the
U.S.?

4. All of the children born to women who were in the Chernobyl area have been normal so far. But will they continue to be healthy? How about future generations?

5. What can be done to deal with the problem of radioactive waste? Can it be disposed of safely? Will it be buried in our backyards?

These are just some of the questions that Americans are asking. Since Three Mile Island, no new orders have been placed for U.S. commercial nuclear reactors. Clearly, Americans fear nuclear power.

But it is also clear that they do not understand nuclear power. The Texas Poll Report of Texas A & M University (Winter 1989) reported that, of Texans who said they "knew something about nuclear power," 38% thought that normally operating nuclear reactors produced air pollution, 50% felt that they were a health hazard for people living nearby, and 55% thought they could explode like an atomic bomb.

None of these perceptions are correct.

But if the public is so fearful of nuclear power, then why have it at all? Isn't it just another dangerous technology that we can do without? The answer is no. We need the electricity. Our present way of producing electricity by burning fossil fuels (coal, oil, and gas) produces respiratory disease, contributes to the greenhouse effect, and depletes the ozone layer. Nor is solar power capable of providing the concentrated energy that we need to run our cities and factories.

Is it just possible that nuclear power is in fact the solution rather than the problem? Chernobyl notwithstanding, can nuclear power actually be a safer alternative to fossil-fuel power? It is important to put nuclear power into perspective and to compare it with other daily risks. For example, the burning of coal in the USSR will cause as much premature cancer *every year* as will be caused in 50 years by the Chernobyl fallout. One must also realize that radiation is a natural part of our environment. Even though 17,000 people may die in the next 50 years of premature cancer due to the Chernobyl accident, many more in this same population will die from exposure to radon, a radioactive gas emitted by the natural rock their homes are built on.

Forty-five years ago, a single atomic bomb was dropped on the Japanese city of Hiroshima. The central region of the city was totally destroyed, and 100,000 people were killed. Three days later, a second bomb fell on Nagasaki. It killed 70,000, and the Japanese immediately surrendered.

Those who were not killed immediately, but who died a few days or weeks later, suffered immensely from burns and radiation sickness. Only 28 of the 300 physicians in Hiroshima remained active after the bombing, and nearly all of the medical supplies were burned, so few of these people could be helped. The world was aghast at the terrible power of these bombs.

The death produced by a single nuclear bomb is fantastically large and far exceeds the death toll that might result from a major nuclear reactor explosion—including the cancers projected to occur late in the lives of the victims. The consequences of a nuclear plant explosion, even one as bad as Chernobyl, do not begin to compare with the devastation produced by a nuclear bomb.

Note also that Hiroshima involved a single nuclear bomb. In contrast, many bombs will be exploded in a nuclear war. These will create extensive smoke and dust that will blot out the Sun, causing unseasonably low temperatures and the consequent loss of food crops and animals. This phenomenon, known as *nuclear winter*, will starve hundreds of millions, perhaps billions, of people, as well as animals and plants, far from the sites of the explosions themselves. And the radioactive fallout from these bombs will induce widespread cancer. Nuclear war will be a thousand times more destructive than any war we have ever known.

Nevertheless, we and the Soviet Union have armed ourselves with tens of thousands of nuclear weapons. A single nuclear submarine now carries warheads with a total explosive force 1000 times that of the Hiroshima bomb—enough to devastate most of the large cities and military bases of either the U.S. or the USSR. In our fear of each other, we have established a system of mutual terror whose implications we seem not to comprehend. The release of these arsenals, far from winning a war, would destroy civilization and possibly all humankind.

Yet we continue to arm. In spite of the epochal changes now occurring in the Soviet empire, President Bush, in his State of the Union message of January 1990, called for reduced conventional forces but no reduction in nuclear forces. In fact, he asked for an *increase* in spending for nuclear war research, including research on the Strategic Defense Initiative (Star Wars). The "nuclear deterrent" is still defended as a reasonable strategy for NATO. In spite of many claims to the contrary, it would seem that the Cold War is not yet over. The world appears to have forgotten the shock of Hiroshima, and to have lost its fear of nuclear war. There seems to be little recognition of the terrible danger of these vast arsenals, and particularly of the foolishness of extending such armaments to outer space.

George Kennan, former U.S. ambassador to the Soviet Union and creator of the policy of containment of the USSR, poignantly described our blindness:

> . . . there is no issue at stake in our political relations with the Soviet Union—no hope, no fear, nothing to which we aspire, nothing we would like to avoid—which could conceivably be worth a nuclear war.[1]

Something is awry here. On the one hand, we see nations arming themselves to the teeth with apocalyptic nuclear weapons and, at the same time, frightened of the hazards of nuclear power. Isn't there something wrong with our perceptions? Is nuclear power really dangerous? Is planning for nuclear war rational? Are we confusing these two applications of the release of nuclear energy, so astoundingly revealed by Einstein in 1905?

This book is addressed to these perceptions, and it attempts to set the record straight. Its writing was motivated by the conviction that there is a great need in America, and indeed in the world, for a better understanding of the benefits and hazards of both nuclear power and nuclear armament.

Now, just a few years from the start of the next millennium, we stand on the threshold of a great new vista. The collapse of the Soviet communist empire provides an unprecedented opportunity for change. We can finally begin to put an end to the frenzy of fear that has led us to build nuclear arsenals that can result only in the end of our world. At the same time, there is a new awareness of the damage we are inflicting on our environment, caused largely by the burning of fossil fuels. This destruction could be averted if the world chose to develop nuclear power in a rational way.

What This Book Is About

Essentially, this book is about Chernobyl and Hiroshima: two monumental events in world history. The book provides a background for understanding their significance. It addresses the perceptions that people have of these events, and it attempts to dispel some of the myths that surround them.

Many people resist learning about nuclear issues. This resistance seems to result largely from three attitudes:

1. The subject is too horrible to think about.
2. The subject is too hard to understand.
3. "Expert" opinions can't be trusted.

Let us examine these perceptions a little more closely.

There are good reasons for the public to be concerned. We are dealing with matters of life and death. With regard to nuclear war, the astronomer Carl Sagan stated it best:

There is no issue more important than the avoidance of nuclear war. Whatever your interests, passions or goals, they and you are threatened fundamentally by the prospect of nuclear war. We have achieved the capability for the certain destruction of our civilization and perhaps of our species as well. I find it incredible that any thinking person would not be concerned in the deepest way about this issue.[2]

But we must go beyond mere concern. We need to conquer our nuclear fear and to move toward nuclear rationality. The press could help us in this endeavor, but too often it simply fuels the public's paranoia. This book should provide some antidote to misinformation; it tries to present a calm and rational view of nuclear power and nuclear armaments.

This is essentially a "background" book in that it attempts to provide not new ideas so much as *an integrated view of existing ideas concerning nuclear issues.* Most of the opinions stated are supported by reference to facts and expert testimony. The book is thus a report of current thinking on these matters and is designed to provide a wholeness of vision.

Nuclear power and nuclear war are complex technological subjects. But they are not beyond the comprehension of the concerned layperson. Some knowledge of the scientific fundamentals of nuclear radiation and nuclear energy* is essential if one is to achieve a true understanding of these subjects. This book provides the necessary basic information in understandable terms. The material in the first section ("Atoms and Life") is highly compressed but it provides important background information for the understanding of later chapters and should be read by the nonscientist. Some experts have advised me against presenting such fundamental scientific information—but I have confidence that concerned laypersons can easily grasp these ideas if they are only willing to try. The book has few equations.

*By **nuclear energy**, I mean the energy produced by atomic nuclei, including its use in nuclear medicine, in the production of electrical power, or in nuclear bombs. I restrict the term **nuclear power** to the production of electricity by nuclear reactors.

Most scientific terms are shown in **boldface** when they first appear in the text, and these are defined in the glossary at the end of the book. There is an extensive index.

A central concern of this book is the nature of the biological effects of nuclear radiations. This book stands alone as a guide for the layperson that follows this thread through the problems of both nuclear power and nuclear weapons.[3]

Finally, part of the problem with trust in "expert" opinions is that the layperson may misunderstand what the experts are really saying. For example, some people believe that the experts called Chernobyl "the accident that could never happen." That is not true. What they said was that such an accident is "extremely unlikely." Catastrophic nuclear accidents *are* very unlikely: we've had only one in over three decades of commercial nuclear power. This book should help laypeople discern what the experts are really saying.

In conclusion, let me make the point that this is not just another scientific book. This is a book about you and your loved ones. It is about matters that concern the future of the human race. It is incumbent on all of us to attempt to solve the complex problems posed for humanity by the release of nuclear energy. If we succeed, future generations will be grateful. If we do not, there may be no future generations.

I   ⚛   Atoms and Life

1 Atoms: What the Universe Is Made Of

Everything, including ourselves, is made of atoms. Nuclear radiations affect humans and other life forms by interacting with these atoms. If we are to understand how radiation acts, we must first know something about the atoms themselves. What are atoms, and how do they behave?

If we look carefully at the material world around us, it appears to be tremendously complex. Even something as simple as a rock becomes complicated when we analyze its composition in detail. A single living cell is extremely complex, and this complexity grows as we ascend to that most advanced community of living cells—the human being—and finally to that marvel of organized cells known as the human brain.

Yet all these complicated structures are made up of only relatively few, very simple things that we call **atoms**. There are 92 different kinds of atom in the natural world, but only about 30 of them occur in the human body.

Some substances, like iron, are made up of only one type of atom. Other substances, like water, are made up of only two different types of atom: hydrogen and oxygen. Dry air contains just a handful of different kinds of atom: oxygen, nitrogen, carbon, and argon.

11

The Physics of Atoms

The atoms, of which everything is made, are exceedingly tiny particles. It takes 200,000,000 of them, laid next to each other, to make a line 1 inch long. How did people ever figure this out? Believe it or not, the idea was first proposed by the ancient Greeks. They decided that water must be made of atoms because otherwise you could not easily move a stick through it. They reasoned that, if water were made of some continuous substance, like a solid, then you could not move the stick at all; and if it were made of, say, long filaments of something, then you could move the stick through it only in one direction: parallel to the filaments. Because you can move a stick in any direction through water *with ease*, then water can only be made of tiny particles that flow around the stick as you move it. With similar reasoning, they extended this idea to all matter. This concept explains why one can drive a nail into a chunk of lead.

Nonscientists sometimes wonder what scientists mean by the "beauty" of scientific reasoning. The Greek atomic theory described above is a classic example: With only pure thought (you can imagine moving the stick through the water—you don't actually have to do it), they were able to arrive at a theory of the structure of matter that was not clearly proved to be correct until 2000 years later.

But what is the atom itself made of? The answer, which could not be discovered until the development of modern science, is startling: *the atom is made mostly of nothing!* All atoms are basically similar. They have a hard and very tiny central kernel, called a **nucleus**, which is so small that its diameter is only 1/10,000 of the diameter of the atom. This nucleus is surrounded by tiny particles called **electrons**, which circle the nucleus at very high speed, somewhat as the planets of the solar system (Earth, Mars, etc.) circle the Sun. They move in orbits that are grouped into a series of spherical "shells" around the nucleus.

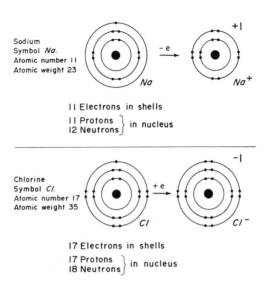

Sodium
Symbol *Na.*
Atomic number 11
Atomic weight 23

11 Electrons in shells
11 Protons ⎫ in nucleus
12 Neutrons ⎭

Chlorine
Symbol *Cl.*
Atomic number 17
Atomic weight 35

17 Electrons in shells
17 Protons ⎫ in nucleus
18 Neutrons ⎭

Figure 1. Electronic structures of the atoms sodium and chlorine and their corresponding ions. The nucleus is actually so small that it would appear only as a tiny dot on this scale.

As in the solar system, all of the space between the electrons and the nucleus of an atom is *empty.* This is one reason why many radiations, like X rays, neutrons, and radio waves, can pass right through your body.

Figure 1 shows a diagram of some atoms. The **mass** (a term from physics equivalent to "substance" or "weight") of the atom is concentrated in the nucleus, which is composed of two kinds of elementary particles, **protons** and **neutrons.*** These two nuclear particles have the same mass and can be thought of as extremely tiny billiard balls. The protons have a positive electric charge and the neutrons have no electric charge (hence their

---

*Except for normal hydrogen, which has no neutrons. Note that all terms defined in the glossary are printed in **boldface** when they first appear in the text.

name). The electrons, which are so light that they can be regarded as having no mass at all, have a negative electric charge. The negative electric charge of the orbiting electron exactly balances the positive charge of the proton in the nucleus. *All normal atoms have equal numbers of protons and electrons and thus have a zero net electric charge.* As seen in Figure 1, the sodium atom has 11 electrons surrounding a nucleus of 11 protons. The atom is thus a tiny electrically balanced world unto itself.

The Chemistry of Atoms

Atoms interact with each other, and the science of these interactions is called *chemistry*. In a solid object like iron, the atoms are stuck tightly together by **chemical bonds**. As the temperature rises, these bonds are weakened, and the iron melts, becoming a liquid. In this state, the atoms can move past each other, although not without *some* resistance. At even higher temperatures, as in the Sun, iron becomes gaseous. In a gas, the atoms are not bonded at all, and the gas flows rapidly into a region of lower pressure, as air does when it comes out of a tire. A gas still shows some resistance to an object, as when an airplane passes through air, but that is because it takes some energy to push the atoms aside: no chemical bonds have to be broken.

Atoms in a gas or a liquid at ordinary temperatures are in very rapid motion. They are constantly hitting each another and bouncing away. Sometimes, however, if atoms collide and happen to be of the right kind, they stick together, in a process we call *forming a chemical bond*. At high temperature, two atoms that are bonded together may be knocked apart by collisions with other atoms, a process called *breaking a chemical bond*. Most of what goes on about us involves only such events, and they make up the world of chemistry. For example, breaking of chemical bonds occurs in a fire, in the gasoline explosions in an automobile engine, and in the breakdown of our body tissues when we exercise. Such breaking of bonds usually releases energy, commonly in the form of heat. Forming chemical bonds, however, usually requires energy, as in building up bone and muscle in an animal, or wood in a tree.

Figure 2 shows a shortened and simplified version of the top part of the **periodic table** of the atoms (often called **elements**). Each atom is represented by a one- or two-letter symbol, like *H* or *Li*. Atoms are arranged in the periodic table according to their properties. In nature, there are 92 different kinds of atoms; only the first 30 are shown here. (The bottom part of the periodic table is shown in Appendix A.) Everything in the world is composed of these 92 atoms or combinations of them.

The various atoms differ only in the number of subparticles that they contain. They range in complexity from the simplest, hydrogen, to the most complex, uranium (not shown in Figure 2). Those atoms to the left of the dark line are defined as *metals*,

$^1_1$H							$^4_2$He
Hydrogen							Helium
$^7_3$Li	$^9_4$Be	$^{11}_5$B	$^{12}_6$C	$^{14}_7$N	$^{16}_8$O	$^{19}_9$F	$^{20}_{10}$Ne
Lithium	Beryllium	Boron	Carbon	Nitrogen	Oxygen	Fluorine	Neon
$^{23}_{11}$Na	$^{24}_{12}$Mg	$^{27}_{13}$Al	$^{28}_{14}$Si	$^{31}_{15}$P	$^{32}_{16}$S	$^{35}_{17}$Cl	$^{40}_{18}$Ar
Sodium	Magnesium	Aluminum	Silicon	Phosphorus	Sulfur	Chlorine	Argon
$^{39}_{19}$K	$^{40}_{20}$Ca	$^{70}_{31}$Ga	$^{73}_{32}$Ge	$^{75}_{33}$As	$^{79}_{34}$Se	$^{79}_{35}$Br	$^{84}_{36}$Kr
Potassium	Calcium	Gallium	Germanium	Arsenic	Selenium	Bromine	Krypton

```
21 - Sc - Scandium
22 - Ti - Titanium
23 - V  - Vanadium
24 - Cr - Chromium
25 - Mn - Manganese
26 - Fe - Iron
27 - Co - Cobalt
28 - Ni - Nickel
29 - Cu - Copper
30 - Zn - Zinc
```

Figure 2. The top part of the periodic table of the atoms (elements). This table includes most of the atoms present in the human body.

many of which are familiar substances, such as iron, written as *Fe*, and aluminum, written as *Al*. Those atoms to the right of the dark line are *nonmetals*, such as oxygen (O) and sulfur (S).

Some things are made of just one kind of atom. For example, a chunk of pure iron is made up only of atoms of iron (although *pure* iron is rare). A chunk of pure sulfur is made only of atoms of sulfur. However, most things in our environment are made of more than just one kind of atom. These are called *chemical compounds*, and they have two or more different kinds of atoms in them.

Some compounds are like table salt, or sodium chloride, which contains equal numbers of sodium atoms that have lost an electron and thus have a net positive charge, written as Na^+, and equal numbers of chlorine atoms that have gained an electron and thus have a net negative charge, written as Cl^- (see Figure 1). Such atoms (called *ions*) in compounds like sodium chloride are held together by *weak chemical bonds*, caused by the attraction of their opposite electric charges. But most compounds, like water, are made of **molecules** and consist of atoms held together by *strong chemical bonds*, in which the outer electron orbits of the bonding atoms actually overlap. A water molecule is composed of two hydrogen atoms bonded to the same oxygen atom by strong bonds. The chemical formula for water is written as H_2O.

Many atoms behave similarly to one another, which is why they are arranged in a "periodic" table of increasing complexity, in which atoms with similar properties lie in the same vertical column. The simplest atom is hydrogen, whose nucleus is a single proton, and it therefore has one negatively charged electron flying around this positively charged proton. The next atom (reading left to right) is helium, which has two protons and two neutrons in its nucleus, and two electrons flying around the nucleus, which balance the two positively charged protons. The next atom is lithium, with 3 protons in its nucleus, surrounded by 3 electrons, and so on. Each subsequent atom in the periodic table has one more proton in its nucleus and therefore one more electron outside the nucleus. The number of protons in an atom's nucleus is indicated by the subscript that precedes each atomic symbol in the periodic table (Figure 2). This is called the **atomic number**, which also represents the number of electrons in a normal atom.

We have noted that the electrons in an atom are grouped into so-called shells surrounding the nucleus (Figure 1). The first, or innermost, shell can contain no more than two electrons (the first *row* of the periodic table), the next shell may have up to 8 electrons (the second row of the table), the next also 8 (the third row), and the next 18. *The chemistry of atoms is determined by the number of electrons in the outermost shell.* The atoms that lie in a single vertical *column* of the periodic table all have similar chemical properties, such as hydrogen (H), lithium (Li), sodium (Na), and potassium (K).* Lithium has two more electrons than hydrogen, sodium has eight more electrons than lithium, and potassium eight more than sodium, but all three have a single electron in their outer shells, and for this reason they have similar chemical properties. For another example, look at the right-hand column in the periodic table. These atoms are the **noble gases**, which are very unreactive chemically. Note that helium (He) has two electrons (subscript 2), neon (Ne) has 10 electrons (8 more than helium), argon (Ar) has 8 more electrons than neon, and krypton (Kr) has 18 more than argon (for a total of 36). All these noble-gas atoms have "filled" outermost electron shells. Atoms with filled electron shells are "satisfied," and they tend not to react with other atoms.

Radioactive noble gases, such as krypton or argon, are sometimes emitted by nuclear power plants. One reason why they are hazardous is that they "float around forever," not combining with atoms in rocks or soil, which would immobilize them. On the other hand, breathing radioactive noble gases is

---

*Some of the symbols for the atoms that appear strange are derived from Latin. Thus the Latin for potassium is *kalium* (K); for sodium *natrium* (Na); for iron *ferrum* (Fe); and for lead *plombum* (Pb) (whence our word *plumber*).

not itself hazardous because they don't react with our tissues, and we breathe them right out again. The radioactive noble gas radon, however, is hazardous, for reasons we shall examine later.

Most of the atoms that make up the world of living things have atomic numbers of 30 or less. They include hydrogen, carbon, nitrogen, oxygen, fluorine, sodium, magnesium, phosphorus, sulfur, chlorine, potassium, calcium, manganese, iron, cobalt, copper, and zinc. Most of the metals present in living things occur in only trace amounts. Human beings are made mostly of water (H_2O) and carbon (C).

Because electrons are essentially weightless, the total weight of an atom is simply the sum of the weights of the protons and the neutrons. Protons and neutrons weigh the same, so the total number of these particles is proportional to the weight of the atom. We call this number the **atomic weight**, and it is shown as a superscript preceding the atomic symbol (see Figure 2). Thus, carbon has 6 protons and 6 neutrons, and therefore has an atomic weight of 12, which is twice the atomic number of 6. As atoms get heavier, they tend to have more neutrons than protons, so that the atomic weight is greater than twice the atomic number. Sodium (Na) has 11 protons, but an atomic weight of 23 (see Figure 1); thus it has an "extra" neutron.

To recapitulate the symbolism used with atoms, we may describe carbon as $^{12}_{6}C$, or as ^{12}C, or as carbon-12. When the name is written out, it is always followed by the atomic weight. The atomic *number* is implied by the name; all carbon atoms, for example, have the atomic number 6, although they may have different atomic weights. Thus, the atom carbon-14 has two more neutrons than carbon-12, but they both have six protons.

Our entire world of normal experience, whether speaking, walking, seeing, eating, growing, or driving a car, is the world of chemistry, and it involves only the *surfaces of atoms* (the electrons in the outermost shell) contacting and reacting with each other. Note, for example, that some of the metals shown in Figure 2 (scandium through zinc) "squeeze" into the periodic table right after calcium. They are atoms in which an internal electron shell is still filling up, but they all have two electrons in their external shell, the same as in calcium. These atoms therefore behave chemically much as calcium does (except for chromium and copper, which have only one outer-shell electron).

Reactions involving the *nuclei of atoms* rarely occur naturally on the Earth and were not artificially produced until quite recently. Such **nuclear reactions** were first revealed to us by scientific experiments performed by Lord Rutherford in 1919. The theory explaining nuclear reactions was developed in the 1920s and 1930s, and culminated in large-scale releases of nuclear energy in reactors and in bombs during World War II.

Why did I say above that we rarely see nuclear reactions "on the Earth"? Because such reactions occur in stars. The region at the center of every star in the sky is basically like a hydrogen bomb, constantly undergoing explosive nuclear reactions. The stars, burning very brightly but relatively slowly, convert hydrogen into helium and other heavier atoms, eventually creating all the other types of atom in the universe. The center of our Sun, a typical star, is such a great hydrogen bomb, continuously exploding and giving off tremendous energy. Even though our Earth intercepts less than one billionth of this energy, the Sun provides all the light and heat that we need to survive. Stars like our Sun will shine for 10 billion years before they burn up their hydrogen fuel.

2 ⚛ Molecules: How the Atoms Fit Together

This chapter is about molecules. We must understand molecules because that is largely what we are made of. Nuclear radiations harm us by damaging our molecules.

Most material things that we encounter in our universe do not have a simple composition. They consist of *mixtures* of things. For example, a stone is usually a mixture of various metals and compounds made of metals, as well as of various nonmetal things, like sand or glass. Chemists can purify these natural substances in a variety of ways, to produce, say, pure water or pure iron. One might think that pure substances like these would have a simple composition. True—and not true. Pure iron consists solely of iron atoms, and is indeed a simple substance. But pure water is composed of molecules, which are groups of atoms firmly bonded together. Molecules may be quite simple, like water, or very complex, like protein molecules.

Why Atoms Get Together

Consider hydrogen gas. It is made of molecules that contain *two* hydrogen atoms. Why? You will remember that the hydrogen atom has only one electron. If two hydrogens get together, or *bond* to each other, they share their electrons, so that each hydrogen nucleus thinks it is surrounded by two electrons instead of one. Why this curious behavior? And why doesn't helium do this? Helium gas consists of *single* atoms of helium.

We noted in the last chapter that atoms like to fill up their electron shells, full shells containing 2, 8, 8, and 18 electrons, as we go up the periodic table (Figure 2). Now let's go back to hydrogen and helium. The helium atom already has two electrons, so it is happy and desires neither more nor less than this number. In fact, this "lack of desire" is what we mean when we say that it is one of the "noble" gases, a gas that shows no tendency to combine with anything else to form a molecule. The isolated hydrogen atom has only one electron, but would like to have two, in order to complete an electron shell. So two hydrogen atoms come close together, permitting their nuclei to share the two electrons. This hydrogen *molecule*, designated H_2, composed of two hydrogen atoms, is now a stable substance with a complete electron shell. Like helium, it has little desire to combine with other atoms or molecules.

Thus, atoms combine with other atoms in an effort to arrive at filled outer shells of electrons, a state in which they are stable.

How Ions Arise

Consider table salt. It is made of sodium atoms and chlorine atoms. If we look at Figure 1 (left side), we see that sodium has a single electron in its outer shell of electrons. One way for it to have a filled outer shell would be to lose this electron. Figure 1 (left side) also shows us that chlorine has 7 electrons in its outer shell. So it is evident that chlorine would like to acquire an extra electron to complete this shell. Now, an interesting thing happens in table salt, or sodium chloride: A sodium atom gives up one of its electrons to a chlorine atom, so they are now both satisfied, both having filled outer shells. Therefore, table salt is a stable compound.

But sodium is *not supposed* to have 10 electrons—this puts it out of electrical balance with its nucleus, which has 11 positively charged protons (Figure 1). This sodium atom thus now has a net electric charge of +1. We call it a *positive ion*. Similarly, the chlorine has now become a *negative ion* (a chloride ion). **Ions**, therefore, are simply electrically charged atoms (Figure 1, right side). They get that way because they seek to have filled outer electron shells or, in other words, to become like a noble gas such as neon or argon.

The sodium chloride crystal is held together by electrical attraction between each positively charged sodium ion and each negatively charged chloride ion. We call this an *ionic bond*. This is a weak bond, and the sodium and chloride ions separate immediately, and remain as ions, when they are dissolved in water. Because these ions are charged, the saltwater solution will now conduct electricity very nicely.

You may be wondering why we are getting so involved in chemistry. There are three reasons: (1) we need to understand what atoms are so that we can understand radioactivity and nuclear fission, (2) we need to understand what ions are so that we can understand how radiations damage molecules, and (3) we need to know about molecules so that we can understand how damage to molecules can hurt us.

Small Organic Molecules

The molecules that concern us most are those of which our bodies are made. These are chiefly *organic molecules*, so called because biological organisms are made of them. All organic molecules contain at least carbon and hydrogen.[1] The simplest organic molecules contain only these atoms and are called *hydrocarbons*. The simplest hydrocarbon is methane, and it has the formula CH_4, meaning that it consists of a carbon atom surrounded by four hydrogen atoms (Figure 3). Why does it have this particular composition? Because carbon has 6 electrons (atomic number 6), two in the inner shell, and four in the outer shell. So carbon wants to have four more electrons in order to fill its outer shell. Carbon therefore associates itself with four hydrogen atoms. The electrons of these atoms are now shared with the carbon, leaving carbon surrounded by the 8 electrons it wants. And each of the four hydrogen atoms gets close to one of

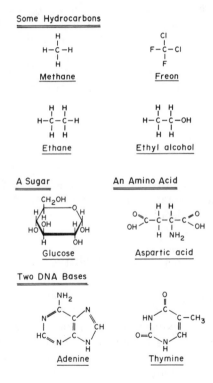

Figure 3. The structures of some simple organic molecules.

the four electrons of the carbon atom, so that each hydrogen is now surrounded by the two electrons it wants. This *sharing of electrons* makes all of the atoms complete, and methane is indeed a fairly stable molecule. It is the major component of the natural gas that many of us use to heat our homes.

The sharing of electrons that occurs in a molecule like methane produces what is called a *covalent bond*. This bond differs from the ionic bond in sodium chloride, where the electrons are *given* to another atom, not shared with it. The covalent bond, indicated by the short lines between atoms in Figure 3, is the most common chemical bond in organic molecules, and it is a strong bond (i.e., not easily broken). Ionic bonds, on the other hand, are weak bonds, but they become important in both the structure and the function of very large molecules, such as proteins and nucleic acids, which possess large numbers of ionic bonds in addition to their many covalent bonds.

Carbon is a remarkable atom. It can produce a multitude of bonds with other atoms, largely because it lies right in the middle of the periodic table, permitting it to combine well with atoms on both sides of the table. In contrast, the metal sodium never combines with the metal magnesium, and the nonmetal oxygen never combines with the nonmetal fluorine. In fact, carbon is so versatile an atom that carbon-containing molecules make up 90% of all the molecules known.

A variety of atoms other than hydrogen can combine with carbon. Let us go back to methane. If you substitute one of the H's with a Cl, you have methyl chloride, a compound produced both biologically and by industry. If you have two Cl's and two F's instead of the four H's, you have Freon, a compound used in refrigerators and air conditioners (Figure 3). Both methyl chloride and Freon are very stable and persist for long times in the atmosphere. Such compounds can damage the ozone layer in the stratosphere, which protects us from dangerous ultraviolet light from the Sun. Thus, both are important environmental pollutants.

Lots of important organic molecules are simple hydrocarbons. In addition to methane, there are ethane (C_2H_6) and propane (C_3H_8), familiar as heating fuels. If one of the H's of methane is replaced by the atomic group OH, you have methyl alcohol. Doing the same with ethane gives ethyl alcohol (C_2H_5OH), the common base of alcoholic liquors (Figure 3).

After hydrocarbons, the next most complex organic molecules are the *carbohydrates*, which are basically *sugars*. These molecules contain oxygen, in addition to carbon and hydrogen. The most fundamental carbohydrate is the sugar glucose, with the formula $C_6H_{12}O_6$. (In Figure 3, five of the six points of the hexagon represent carbon atoms.) Glucose is a component of sucrose, our common table sugar. Glucose is the primary energy-supplying molecule in biological organisms, providing energy for muscle and nerve action, as well as a host of other vital functions.

Large Organic Molecules

Now we come to the really fascinating molecules that are produced only by living systems, namely, the polysaccharides, proteins, and nucleic acids, all of which are very large molecules. These large molecules are called *polymers*, because they are made of chains of similar or identical smaller molecules, or building blocks, which are called *monomers*. Glucose is stored by plants as a long chain of glucose molecules called starch, and by animals as a long chain called glycogen, which is stored particularly in muscles. Thus, both starch and glycogen are *polysaccharides*,* both being polymers of glucose, which is a sugar, or "saccharide." When we eat vegetables, we break down starch to glucose, and when we eat meat (muscle), we break down

---

*In common usage, both the polysaccharides, such as starch and glycogen, and the simple sugars, such as glucose and sucrose, are referred to as carbohydrates. Sucrose, or "table sugar," is a *di*saccharide, consisting of a glucose molecule bonded to a molecule of fructose, or "fruit sugar."

glycogen to glucose. This glucose provides energy for our bodily processes.

Proteins are the most widespread, complex, and versatile of the large molecules. Like the other large molecules, they are polymers. The building blocks (or monomers) of which they are built are called *amino acids*. The amino acid, aspartic acid, shown in Figure 3, has some "double" covalent bonds, indicated by double lines, which are very strong bonds. The amino acids are small molecules, about the size of a sugar, but there are 20 different kinds of them, and *their sequence in the polymer is specific*. Proteins of the same type have identical sequences of amino acids. Furthermore, every protein of the same type has an identical length, usually about 100 amino acids. This situation is quite different from that of a polymer such as starch, which has only one kind of monomer and may have any length. The specific amino-acid sequence and length endow each protein with unique properties. Our bodies contain tens of thousands of different proteins, but each particular protein, such as the enzyme called amylase, in our saliva, is identical to every other molecule of that protein.

Amino acids do not always go into making proteins. Many of the *neurotransmitter* molecules that cause our nerves to function, as well as many hormones, such as vasopressin, which raises blood pressure, and antibiotics, such as gramicidin, are short chains of just a few amino acids. The artificial sweetener aspartame (trade name, NutraSweet) is a compound made up of only two of the amino acids, aspartic acid and phenylalanine.

Proteins are of two fundamental kinds. One is *fibrous protein*, which consists of long protein chains twisted around one another, much as cotton thread is made up of cotton fibers. One fibrous protein is keratin, which composes fingernails, hair, horn, and beak, as well as the fibers that give our skin its toughness. Another important fibrous protein is collagen, which composes the connective tissue of our soft body parts, and makes scars and tendons. Collagen makes up 25% of all animal protein.

The other basic kind of protein is *globular protein*. These molecules are roughly spherical and are made of a long chain of about 100 amino acids that coils and twists around itself to form a sort of ball, something like a ball of twine. This ball has a very specific shape. Every bump and crevice on its surface is exactly the same in every protein of the same type. Such globular proteins compose the **enzymes** of living tissues.

Now, the enzymes are of extreme importance, as they make everything in our bodies. A special crevice in the enzyme molecule can react very rapidly and very specifically with a particular small molecule (called a *substrate* molecule) with which it forms a tight fit, and this tight fit gives such proteins a very particular chemical specificity. This reaction specificity is of great importance in biological systems: it enables a specific protein to produce a specific chemical reaction, such as breaking a bond in a certain molecule. Because an enzyme interacts very rapidly and specifically with such a small molecule, it causes it to undergo a specific chemical reaction. The enzyme binds to the substrate molecule by several ionic (weak) bonds. The enzyme being much bigger than the substrate molecule, this binding does not deform the enzyme, but it does deform the substrate, permitting a single atom or a small group of atoms in the substrate to be more easily detached by the normal collisions with other atoms or molecules. Thus, an internal covalent (strong) bond in the substrate may break, or a new covalent bond may form. The enzyme is therefore a *catalyst* for that particular reaction. For example, when we chew some toast, the digestive enzyme amylase in our saliva breaks off glucose (sugar) monomers from the starch polymer (the substrate) in the toast. This is why toast tastes sweet in our mouths.

We now come to the third and last kind of large molecule: *nucleic acids*. These have the long names deoxy-ribo-nucleic acid (**DNA**) and ribo-nucleic acid (RNA). We shall discuss only DNA. Like the other large molecules, DNA is a polymer. Its basic unit, or monomer, is called a *nucleotide*, which consists of a sugar called ribose (whence the *ribo* in the names), a phosphate group (PO_4), and a small molecule called a *base*. Nucleotides all have similar structures except for the bases, of which there are four kinds in DNA: adenine (*A*), thymine (*T*), guanine (*G*), and cytosine (*C*) (see Figure 3).

The sugar and phosphate groups of DNA form a long chain, called the *backbone*. The bases stick out from this structure at right angles. A complete DNA molecule consists of two such long chains twisted around each other to form what is called a *double helix*. This is shown in Figure 4, where the dark pentagons are the ribose sugars, and the light lines connecting them are the phosphate groups. The bases, which are flat molecules seen edge-on, project toward each other from the two backbones.

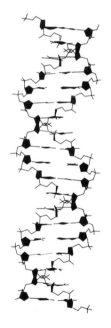

Figure 4. The double-helix DNA molecule. In this form, the DNA is inactive. (From Lubert Stryer, *Biochemistry*. Copyright 1975 by Lubert Stryer. Reprinted with permission by W. H. Freeman & Company.)

The bases from opposite chains are held together by very weak ionic bonds, called *hydrogen bonds* (not shown in the figure).

Genetics and Metabolism

Virtually every chemical reaction that goes on in our bodies is catalyzed by a particular enzyme. What our living tissues do is chemistry. Our stomach and intestines, using digestive enzymes, break down our food into its building blocks, such as amino acids and sugars. Our cells then build up their own proteins and polysaccharides from these building blocks. In this way, they make us grow and store energy. Our living tissues also break down molecules to give us available energy for running, breathing, and so on. All this building up and breaking down, called *metabolism*, that goes on in our bodies requires a large number of different chemical reactions. For example, using the oxygen we breathe into our lungs, about 40 different reactions break down a glucose molecule completely to carbon dioxide and water, which we breathe out from our lungs. In this process, we get a very large amount of energy from the glucose. Every one of these 40 reactions is mediated by a different protein enzyme. That every chemical reaction in our bodies is mediated by an enzyme has far-reaching consequences. It means that, if we have all the enzymes we need, then our bodies can make or build up anything that they require, including all our body structures, such as bone and muscle. Our bodies can also break down any tissue to get energy from it, which happens when we starve. So, to build a biological system, one just has to specify what all the enzymes are going to be.[2] The process of specifying the enzymes is what *genetics* is all about.

DNA is the material that specifies the enzymes. It is a very long and thin, threadlike molecule, containing a linear arrangement of the four bases, *A, T, G,* and *C*. These bases are arranged in a specific sequence that carries a message. *This message specifies the amino-acid sequence of a protein.* The message is written in three-letter words, called the **genetic code**, and the alphabet has only four letters—the four bases: *A*, *T*, *G*, and *C*. Thus, it is a very simple language, but its three-letter words have enough different spellings, theoretically 64, so that it can easily specify the 20 different amino acids that make up proteins. For example, the genetic code in the DNA for the amino acid aspartic acid is *CTG*. Every time the base sequence *CTG* occurs in the DNA, the amino acid aspartic acid is inserted into the protein that is being made. This is a very complex process, and involves a local spreading apart of the DNA double helix—easily done because the hydrogen bonds are weak—so that enzymes can read the "information" on one of the strands. The closed double helix shown in Figure 4 is actually the inactive form of DNA.

One of the most important findings of 20th-century biology is that the function of the genetic material (DNA) is to specify the amino-acid sequence of proteins, thus producing a long chain of amino acids. This long chain will later *spontaneously* fold into the proper three-dimensional form for that protein. In other words, the three-dimensional form is determined by the amino-acid sequence. Most of the proteins that are made by the DNA are enzymes.

That *segment of the DNA chain* that specifies the amino acids for one particular protein is called the **gene** for that protein. All the genes together (or all the DNA) are called the *genetic material* of a living system. Virtually the only function of genes is to make proteins. Thus, each DNA molecule is a very long chain of genes, which specify the structure of thousands of different proteins. If you can specify all the proteins in a living cell, you are specifying all the enzymes, which in turn can make all the other molecules in the cell.

Cells

Our bodies are made of soft tissues, like liver and muscle, and hard tissues, like hair and bone. Soft tissues are made up of tiny living **cells**, which are globs of very complex fluid surrounded by a "cell membrane." These cells have a little sphere inside them called the **nucleus**; all the rest of the interior of the cell is called *cytoplasm*. There may also be some cell products, such as the collagen fibers in connective tissue, that can run through the cells as well as outside them, thus connecting them to other cells. Hard tissues are made up mostly of cell products as well as the remains of dead cells, as in hair and bone.

In living cells, the DNA resides in molecular structures called **chromosomes** (the name comes from the fact that we can see them in a microscope if we color them with a dye), which are located within the cell nucleus. In human cells, there are 23 different chromosomes that came from the mother and 23 similar (but not identical) chromosomes that came from the father. Each chromosome contains a single, very long molecule of DNA. This single molecule contains thousands of genes, each gene being simply a segment of the DNA. Each gene on the DNA of a paternal chromosome is functionally the same as that on the maternal chromosome, but some differ slightly, causing the cell to resemble the corresponding cell of either the mother or the father. Thus, all human babies have the characteristics of humans (rather than apes or dogs), but they resemble one parent or the other in certain fine points, such as the shape of the nose or the color of the hair.

Because DNA has the information for making everything in the cell, it is not hard to see that damage to a single DNA molecule can be disastrous for the cell. Radiation can not only damage the bases in a DNA molecule but it can also break the sugarphosphate backbone, thus causing the chromosome to break. Such damage can activate genes that were dormant, leading to excessive cell growth, or cancer. However, if enough cells are damaged, as in people exposed to a nuclear explosion, the tissue made up by these cells (such as the intestinal lining or the bone marrow) may cease to function in a normal way, and the victim may die.

Let us try to organize the complex information that we have

presented about molecules and genes. We live by eating food, which mostly contains the large molecules of carbohydrates, proteins, and nucleic acids that are characteristic of the plant or animal we are eating. This food is broken down in our digestive tract by proteins, called digestive enzymes, into the basic building blocks of the large molecules: sugars, amino acids, and bases. These building blocks then enter our bloodstream through the intestinal wall, travel to the cells of various tissues, and are there built up, by more enzymes, into new polysaccharides, proteins, and nucleic acids characteristic of our own bodies. If we want to build a new body structure, like some muscle or skin, then the enzymes in the cells use the building blocks to perform this task. All of these processes require energy.

On the other hand, if we *need* energy, then the enzymes in the cells attack the building blocks, breaking them down to very simple molecules like carbon dioxide and water. During this process, the enzymes extract energy from the chemical bonds of the molecules. This activity requires oxygen, which we get by breathing air into our lungs, where the oxygen diffuses into the blood capillaries. All of this metabolism taking place in our tissues is carried out by enzymes. The enzymes have very specific structures, so that each enzyme is optimally designed to carry out its own specific reaction and no other. As the cell is growing and producing enzymes, the structure of the enzymes is determined by the sequence of bases in the genetic material, DNA.

These processes go on in all of the living cells that make up the tissues of our bodies. The part of this system that is most vulnerable to damage, including the damage produced by radiation, is the DNA, because it contains all the genes—the "master plan"— for all the proteins in living cells.

Finally, the DNA structure is *inherited*, so that daughter cells have DNA identical to that of parental cells. Genes are thus passed on from cell generation to cell generation. Damage to a cell may therefore affect all the progeny cells derived from it. If such cells are our germ cells, then the damage will be passed on to their progeny cells—which become our children.

3 Radiation: How the Atoms Interact

"Radiation" has become a horror word for most Americans. Comic-strip fiction is full of exotic and dangerous radiations, and every hero of the future has a ray gun. In the real world, we hear about deadly radiation from nuclear power plants and nuclear bombs, and we know that X rays can penetrate our bodies and damage human embryos. It certainly doesn't sound like very nice stuff.

But in fact, radiation is the very stuff of life—and of the universe. Physicists see the universe as being composed of only two things: matter and radiation. As we have seen, all matter is composed of atoms, and in most forms of matter, several different atoms are grouped together to form molecules. There are also subatomic *particles*, such as protons and neutrons, which are components of atoms. The whole material world, therefore, can be said to consist only of atoms, groups of atoms (molecules), or parts of atoms (particles).

But atoms by themselves cannot do anything. Without some external force, they would just sit there, as the atoms in interstellar dust have sat in outer space for billions of years, or as the atoms in the oldest rocks on Earth, like the granites in the Canadian Shield, have just been there, doing nothing, for billions of years. In order to do something, the atoms must interact. They do this by means of radiation.

There are more kinds of radiation than most people are aware of. Gravity, for example, is actually a form of radiation. We are concerned chiefly with the kind of radiation that is very familiar to us, namely, light. Light consists of tiny entities called **photons**, which have zero mass, meaning that they weigh nothing. But these units of light compose far more than just the visible light that we detect with our eyes. X rays and radio waves exist as photons, for example. Photons also mediate what we call the *electromagnetic force*, which is the force exerted between molecules when we strike a ball with a bat: the molecules of the ball never actually touch the molecules of the bat—they come very close together, and then what physicists call an "exchange of photons" produces an electromagnetic force that pushes them apart. Because these photons have electric and magnetic properties, they are called **electromagnetic radiation**.

Electromagnetic Radiation

We are all familiar with light, but we are usually not aware of the fact that it consists of photons. When you see something in your environment, like a ball, it is because photons of light from the sun strike the ball, bounce off it, and travel to your eye, where they are detected by the retina and converted into an electrical signal that travels to your brain. At night, when the sun's photons are not available, you can't see the ball, unless you provide your own source of photons, such as a flashlight.

The light with which we are most familiar is visible light. But visible light is just a small part of a vast spectrum of light, *most of which we cannot see or feel*. Photons of light carry **energy**, which is a measure of their ability to do work. Different kinds of light have different energy; they can therefore be arranged on a scale of energy that we call the *electromagnetic spectrum*. We see in Figure 5 that this spectrum ranges from the extremely energetic **gamma rays** and **X rays**, through the moderately energetic ultra-

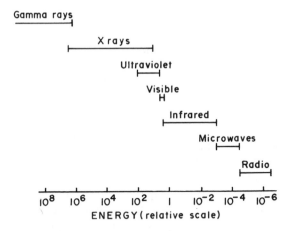

Figure 5. Diagram of the electromagnetic spectrum. Note the tiny portion occupied by the visible-light region.

violet light, visible light, and infrared light, to radiations of very low energy, the microwaves and radio waves. Note that the energy scale in Figure 5 is logarithmic (i.e., each "tic" represents a factor of 100), so a gamma-ray photon has roughly one million (10^6) times more energy than a photon of visible light.

Does this mean that an X ray is the same thing as visible light or a radio wave? The amazing answer is yes: they are all photons of light that simply have different amounts of energy. Our eyes are able to see only the photons of visible light, but some creatures can see light that we cannot see. Many insects, such as bees, do not see red but do see ultraviolet light. This is why we use red or yellow lights to avoid attracting bugs, but violet or ultraviolet lights to attract them to a trap.

The electromagnetic radiations that will concern us are the X rays and the gamma rays, since they are the most energetic, and therefore the most dangerous.

Particle Radiation and Radioactivity

A radiation is simply something that radiates out from a center. Thus we say that a fire "radiates heat," and we may even say that a person "radiates enthusiasm." Because this usage of the term *radiation* is so broad, it is not surprising that scientists distinguish two quite different kinds of radiation.

We have already discussed electromagnetic radiation. The other kind is called **particle radiation**. This consists of parts of atoms, called **subatomic particles**. They include the normal constituents of atoms—protons, neutrons, and electrons—as well as a more unusual particle, called an alpha particle. These particles have mass—unlike photons, which have no mass.

Before discussing particle radiations, we must consider the heaviest atoms that exist. There are 92 naturally occurring atoms, going from hydrogen to uranium; Appendix A shows the heaviest ones, at the bottom part of the periodic table. There are also some manmade **transuranic atoms**, such as neptunium and **plutonium**. Transuranic means "beyond uranium"—such elements are heavier (have a higher atomic number) than uranium. Transuranic atoms are made only in nuclear reactors or particle accelerators, machines that can bombard the nuclei of uranium or other heavy atoms with neutrons or protons, causing these nuclei to be converted to a heavier form.

To simplify matters, the atomic weights (superscripts) shown in Figure 2 and Appendix A are whole numbers. But the real atomic weights are not whole numbers. For example, the atomic weight of carbon is not exactly 12—it is 12.01. The reason is that, in nature, there are different forms, or **isotopes**, of carbon. Most carbon atoms have 6 protons and 6 neutrons, giving an atomic weight of 12. However, about 1% of carbon atoms have 6 protons and 7 neutrons, giving an atomic weight of 13. The average atomic weight of natural carbon is therefore slightly greater than 12, or 12.01. All atoms have several different normal isotopes, and therefore none of the average atomic weights are whole numbers. Note that *all* atoms are isotopes, not just those that differ from the most common form. Thus, both carbon-12, the common form, and carbon-13 are different isotopes of carbon.

All isotopes of a given element have the same number of protons, and therefore behave chemically in the same way. This is because the chemistry depends upon the number of electrons in a neutral atom, which in turn is equal to the number of protons.

As noted in Chapter 1, we rarely see reactions that involve the nucleus of the atom, chemistry being concerned with the outer electrons. We first became aware of the possibility of nuclear reactions at the end of the last century, when "radioactivity" was discovered. What is radioactivity?

Radioactivity *is the emission of radiation from the nucleus of an atom.* The radiation may be particulate or electromagnetic. This astounding process was not understood when it was first investigated by such people as the great French physical chemist Madame Curie. But we now understand it quite well.

The ratio of neutrons to protons is usually 1.0 in light atoms, but it slowly rises above 1.0 as the atomic weight increases. Thus, helium has two protons and two neutrons in its nucleus, and carbon has six protons and six neutrons, but iron has 26 protons and 30 neutrons. The normal isotope of uranium, the heaviest atom, with an atomic weight of 238, has 92 protons and 146 neutrons (238 − 92 = 146).

If the ratio of neutrons to protons in the nucleus of an isotope is sufficiently great, the nucleus may be unstable. Such a nucleus will achieve stability by *spontaneously* changing its composition, in a slight way. It will do this by emitting an energetic (fast-moving) particle, consequently *converting itself into the nucleus of an entirely different atom*, called a **decay product** or a **daughter isotope**. This process, of radiation coming from a nucleus, is called *radioactivity*. Atoms that do this are called *radioactive isotopes* or simply **radioisotopes**.

It may take only seconds for a particular atom to "decay" in this fashion, or it may take many years. Thus, in a rock that contains a radioactive isotope, a few of the atoms will decay in the first minutes, but most will take longer times, and some may even take many years to decay. Figure 6 shows that the radioactivity of a sample—in this case radon gas—decreases rapidly to start with, but then the activity slows down as time goes on and more and more of the radioisotope atoms have decayed—until eventually the radioactivity is essentially zero. However long it takes for one-half of the atoms in a chunk of matter to decay is called the **half-life** of the reaction. The half-life of radon-222 is 3.8 days. It is important to recognize that it is only the *average* atom of radon that will decay in 3.8 days: some radon atoms will decay in the first second, while others may not decay until after several weeks have passed. After 19 days—5 half-lives—the radioactivity of radon is only 3% of its initial value.

Often the daughter isotope produced by radioactive decay

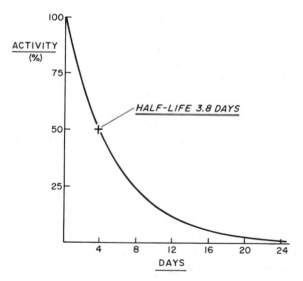

Figure 6. Radioactive decay of radon-222. After 1 half-life (3.8 days), the activity has fallen to one-half of its initial value. After 5 half-lives (19 days), the activity is down to 3% of its initial value.

is itself radioactive. In this event, the daughter atom will also decay, and so on, until a stable nonradioactive nucleus is achieved. Thus, radioisotopes of very high atomic weight go through a whole series of decays from one product to another, before reaching a stable end-product. Figure 7 shows that uranium-238 (^{238}U) decays all the way down to lead-206 (^{206}Pb), which is stable. The half-lives of these emissions are shown in the circles. The decay of uranium-238 to thorium-234 has a half-life of 4.5 billion years, but the half-life for decay of radon-222 to polonium-218 is only 3.8 days.

The 4.5-billion-year half-life of uranium-238 (^{238}U) is about the same as the age of the Earth. By measuring the relative amounts of ^{238}U and its decay products in a mineral, such as a zircon, we can tell how long the mineral has existed in solid form. This is how we know that the solid Earth is 3.8 billion years old. This immensely useful technique is called *radioactive dating*.

Let us examine the decay scheme of uranium-238 more closely. The first reaction is a decay of uranium-238 ($^{238}_{92}$U) to thorium-234 ($^{234}_{90}$Th). This reaction involves the emission of an **alpha particle** from the uranium nucleus. An alpha particle is the same as a helium nucleus, which consists of two neutrons and two protons, so it has a double positive electrical charge. If the uranium atom loses an alpha particle, it loses two protons, and its atomic number will drop from 92, which is uranium, to 90, which is thorium. Also, the uranium nucleus loses four units of atomic weight (2 protons and 2 neutrons), so that the atomic weight is lowered from 238 to 234. Every time we have an alpha emission, there will be a drop of 2 units in atomic number and 4 units in atomic weight. This process is represented in Figure 7 by arrows pointing downward to the left. The figure shows that there are nine alpha emissions in the decay scheme of uranium-238. All uranium-238 atoms will decay in this way, but the overall process takes a very long time.

In addition to the emission of alpha particles, decaying uranium atoms also emit electrons, called **beta particles**, from the

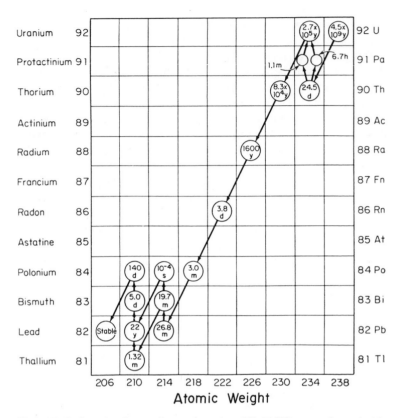

Figure 7. Radioactive decay scheme of uranium-238. Half-lives are shown inside the circles, where s = seconds, m = minutes, d = days, and y = years. (From Ralph E. Lapp and Howard L. Andrews, *Nuclear Radiation Physics.* Copyright 1963, p. 73. Reproduced by permission of Prentice-Hall, Englewood Cliffs, New Jersey.)

nucleus. But how is this possible, as the nucleus has no electrons? Remember that a neutron has the same weight as a proton but does not have an electric charge. Physicists have learned the remarkable fact that a neutron behaves as if it consisted of a proton (with a + charge) combined with an electron (with a −

charge). This fact makes sense, because the electron has essentially no weight, and the positive and negative electric charges cancel each other. Sometimes, in an atomic nucleus, *a neutron decays into a proton and an electron*. When this remarkable event occurs, the electron is emitted, while the proton stays in the nucleus. The emitted electron is called a *beta particle* or *beta ray*. When a nucleus undergoes such a *beta decay*, there will be no change in atomic weight, because only an electron is lost. But there will be an *increase* of one unit in atomic number, because a new proton has been created in the nucleus. In the radioactive decay of uranium-238 shown in Figure 7, the vertical arrows represent such beta decays. For example, after the alpha decay of $^{238}_{92}U$ to $^{234}_{90}Th$, the thorium undergoes a beta decay to protactinium-234 ($^{234}_{91}Pa$). This decay can occur in two different ways, each with a different half-life, indicated by the two upward arrows. In the overall decay scheme of uranium-238, there are 9 beta decays.

Radioisotopes emit either an alpha particle or a beta particle, or sometimes both. In addition, these radioactive decays are usually accompanied by the emission of a gamma ray, which is high-energy electromagnetic radiation. This gamma emission occurs because a nucleus that has decayed by particle emission is left in a state of high energy; the nucleus loses this extra energy by emitting a gamma ray.

In summary, there are three radiations that may be emitted by radioisotopes: alpha rays, beta rays, and gamma rays.

Many of the daughter isotopes shown in Figure 7 have quite short half-lives. They will never be present in great amounts because, as soon as they are created, they decay again to something else. Only those with long half-lives, such as uranium-234, thorium-230, radium-226, and lead-210, will be present in large amounts in a rock containing uranium.

However, radon gas ($^{222}_{86}Rn$), even though it has a half-life of only 4 days, is dangerous. It can build up, from the decay of radium, to dangerous levels in coal mines and uranium mines. The danger comes not from the radon itself (which as a noble gas does not react with our tissues and delivers only a small dose of radiation to the lungs) but from its daughter radioisotopes—polonium, bismuth, and lead—which are solids that combine with small particles in room air. When inhaled, these particles remain embedded in the lung tissue, where their alpha particles can produce lung cancer. Radon gas is the immediate daughter of radium, which is present in many rocks (typically granites) at the surface of the Earth. A home built on such rock can build up dangerous levels of radon. Many homes in the Rocky Mountains and in an area called the Reading Prong (Figure 10, Chapter 4) contain high levels of radon, in some cases sufficient to increase considerably the chance of lung cancer. This problem has only recently been recognized.

Absorption of Radiation

In this book, we are concerned only with high-energy, or "energetic" radiations. These radiations are emitted by nuclear bombs or by the "fallout" from a nuclear power plant explosion, and they are especially dangerous for living things. Thus, of the various types of electromagnetic radiation shown in Figure 5, we shall deal only with X rays and gamma rays. We shall also deal with particle radiation, most of which is energetic.

Radiations harm us only by interacting with our body tissues or, as we say, by being "absorbed" by our tissues. We must therefore examine how this interaction occurs.

1. *Charged particles* (*electrons, protons, and alpha particles*). When traversing matter, electrically charged particles occasionally hit an atom. When this happens, they knock an electron out of the atom, producing an *ion pair*, which is composed of the negatively charged electron that is ejected, plus the positively charged atom (ion) that is left behind. Such events are called *ionizations*, and we therefore speak of energetic radiation as **ionizing radiation**.

A charged particle typically produces thousands of ionizations as it travels through matter. Like a billiard ball hitting other balls, with each collision that produces an ionization it loses a little energy. Eventually, it has lost all its energy and then it stops, and the total distance it has traveled before it stops we call its "range." *Charged particles, such as electrons, protons, and alpha particles, have short ranges.* A typical electron, or beta ray, will penetrate human tissue for only a couple of millimeters and can be stopped by a sheet of glass. An alpha particle has even less ability to penetrate matter and is stopped by a piece of thick paper. *These radiations are dangerous only if the atoms emitting them are taken inside the body, by eating, drinking, or breathing.* Thus, radioactive strontium (^{90}Sr), which emits only beta rays, and behaves chemically like calcium, is absorbed by bone, but it damages only the bone tissue immediately adjacent to it, because its beta rays have a very short range. Such a radioisotope outside the body will not be harmful, even if it falls on the skin. The weak beta radiation does not penetrate far into the skin, and the isotope is soon washed or worn off.

2. *Neutrons.* Neutrons also eventually produce ions. They lose energy in matter chiefly by colliding with protons and transferring energy to them, thus creating energetic (rapidly moving) protons, which in turn lose energy through ionization. Good neutron absorbers include water, whose hydrogen nuclei are simply protons, and concrete, which contains a lot of water molecules and is therefore largely hydrogen. Neutrons are well absorbed by human tissue, which is largely water. They are also absorbed well by other atoms of low atomic weight (light atoms), such as the boron used in nuclear reactor control rods, and the graphite, or carbon, used to slow neutrons down in nuclear reactors. Most neutrons have high energy and penetrate deeply into human tissue before losing all their energy. *A high-energy neutron typically has a range of about 10 centimeters in human tissue.* Less energetic neutrons do not penetrate very far, whereas more energetic ones may go right through us without losing much energy by collision with protons. The gruesome coincidence that neutrons produced by nuclear bombs usually have optimal

energies for absorption in a human body is the basis for the effectiveness of the neutron bomb.

3. *Electromagnetic radiations (X rays and gamma rays)*. X rays and gamma rays behave rather differently from the particles. When a photon of such radiation strikes an atom, it ejects an electron from the atom, thus ionizing it, and producing an energetic electron. These electrons, just like beta rays, lose energy in matter by producing many further ionizations. However, unlike a charged particle, *the X ray or gamma ray may penetrate deeply into tissue before it loses all of its energy through ionization.* Like the neutrons, the gamma rays emitted by radioisotopes and by nuclear reactors and bombs generally have just sufficient energy to pass through our bodies, and they lose most of their energy in the process. Gamma rays are absorbed best by atoms of high atomic weight (heavy atoms), such as lead or iron.

4. *Summary.* Let us now recapitulate. *The end result of the absorption of any energetic radiation is the production of ionizations.* Ionizations easily break chemical bonds. They thus destroy molecules by breaking them up. One ionization in an enzyme (protein) or a gene (DNA) will destroy its biological function. A single ionization in a gene can kill a living cell. This is why energetic radiations are so damaging to life.

Some radiations are much more dangerous to humans than others. Neutrons and gamma rays are very dangerous because they penetrate deeply into our bodies and lose almost all of their energy on the way. Charged particles, like alpha particles and beta rays (electrons), do not penetrate deeply and are dangerous only if the atoms producing them are ingested or inhaled. Radioisotopes that emit only alpha or beta rays, with no gamma rays, such as polonium-210 (see Figure 7), pose little danger if they fall on the skin, although they can damage the lips or eyes. However, some beta-emitters, such as thallium-210, do emit energetic gamma rays and are dangerous if they remain very long on the skin.

With this knowledge of how various radiations are absorbed in matter, we can decide how to protect ourselves. The charged particles—like protons, alpha particles, and beta rays—are easily shielded against—heavy clothing will often suffice, and the face can be protected with a glass face mask. The neutron, being uncharged, is harder to shield against, light atoms being the most effective absorbers. Good readily available absorbers of neutrons are water and concrete. Thus, people inside concrete buildings are well shielded against neutrons coming from outside. Spent fuel rods from nuclear reactors are usually stored under water. The control rods of nuclear reactors are often made of the chemical element boron because it absorbs neutrons so well. Tanks of water containing boron may be used to "douse" the reactor core in an emergency.

X rays and gamma rays are the most difficult to shield against; the best absorbers are heavy atoms, like iron or lead. For this reason, lead aprons are worn by X-ray technicians and are also used to cover parts of the body that one doesn't want to irradiate during X-ray therapy. Soldiers inside steel tanks are well protected against gamma rays but not against neutrons, which is part of the rationale for the use of neutron bombs on the battlefield.

A final point about radiation. It is important to note that the intensity of all radiations falls off as the distance from the source *squared*. In other words, if one is 2 feet away from a radiation source, the intensity of the radiation will be only $\frac{1}{4}$ of that at 1 foot ($2^2 = 4$), and at 3 feet the radiation will be only $\frac{1}{9}$ as intense as at 1 foot ($3^2 = 9$). This means that a very intense radiation source may not be at all dangerous if one is a mile away from it. In addition, although air is not very dense, and therefore a foot of air is not a very good absorber, a mile of air will absorb a lot of radiation. These considerations are very relevant to the danger of radiation from a nuclear power plant or from an exploding nuclear bomb.

In these first three chapters, we have learned about the atoms and molecules of which we are made. We have also studied the nature of radiation and how high-energy radiations are absorbed by atoms and molecules, producing ionization. We now move on to a discussion in some depth of the effects of high-energy radiations on humans, and of how such radiations can be used in medicine and industry to promote human welfare.

II ⚛ Radiations and Life

4 Radiation Biology

Ionizations and DNA

We have seen that the energetic radiations that concern us are all *ionizing radiations*. This means that all the biological effects of energetic radiations are due to the ionizations that they produce, whether they are particle radiation, like protons or neutrons, or electromagnetic radiation, like X rays or gamma rays. The science of the actions of ionizing radiations on biological tissue is called *radiation biology*. It dates back to 1899, when X rays were first used to treat cancer.

The ionization of an atom in a molecule almost always results in the breaking of a covalent (strong) chemical bond. This means that the molecule that suffers an ionization is so badly damaged that it loses much of its normal function. This is true regardless of the role played by the molecule in a cell. The ionizations produced by radiation occur at random throughout the living cell. Therefore, all molecules are vulnerable, and we have to decide which ones are most important to cell function.

Many enzymes are critically important to the cell. However, each particular enzyme is present in hundreds or even thousands of copies, so if some of them are damaged, there are always others to take over. Furthermore, if the gene that makes the enzyme is intact, then the cell can make more of that enzyme at any time. This is true as well of all the other molecules in the cell, except DNA. DNA is generally present in only two copies per cell, so the cell has only two copies of each gene. Damaging even one of these genes can be devastating to cell function, because some critical protein may no longer be produced. Clearly, then, the most crucial molecule in a cell is DNA: the genetic material. If you damage the DNA, you damage the "master plans" that determine what the cells will make and do. *The DNA of living cells is the most important molecule that is damaged by ionizing radiations.*

Figure 8 shows how ionizing radiation damages DNA. Part of the damage—called *direct action*—is caused by ionizations produced by the radiation directly in the DNA molecule itself. Another part—called *indirect action*—is caused by ionizations that the radiation produces in surrounding water molecules. Such ionizations break the water molecule apart, releasing electrons and *free radicals*, which are forms of atoms or small molecules that are highly reactive. Free radicals have no electric charge and thus differ from ions. The most important one is the OH radical, which is an oxygen atom bound to a single hydrogen atom. The electrons and OH radicals migrate to the DNA, react with it, and damage it. Either direct or indirect action can produce the three DNA damages shown in Figure 8 as (1) a double-strand break in the DNA (a break in both of the sugar-phosphate backbones), (2) the deletion of a base, or (3) the cross-linking of the two DNA strands by the production of a new covalent bond between bases on different strands. Any of these kinds of damage alter the ability of that part of the DNA to perform its genetic function of coding for a protein.

We call any such alteration in DNA a **mutation**. Mutations are produced not only by ionizing radiation, but also by certain chemicals in our food, such as the aflatoxins that exist naturally in peanuts. Mutagenic chemicals, such as those in coal smoke, are also present in the air we breathe.

All the cells in our bodies, except those concerned with reproduction, are called **somatic cells**. Damage to their DNA may kill these cells, and we would call this damage a **lethal**

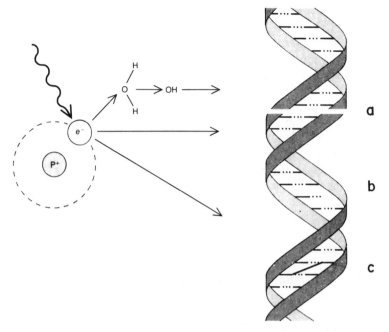

Figure 8. Radiation damage to the DNA molecule. (From A. C. Upton, "The Biological Effects of Low-level Ionizing Radiation." Copyright 1982 by SCIENTIFIC AMERICAN, Inc. All rights reserved.)

mutation. It is this lethal action of radiations that is used in *radiation therapy* to destroy cancer cells.

However, the cell that has been irradiated may suffer a **nonlethal mutation**. What then? The cell goes on living, but it may have lost the ability to do something. In most cases, this is no problem because there are many other cells that can take over. Occasionally, however, the nonlethal mutation is one that affects the control of cell growth. A cell with such a mutation may start dividing uncontrollably, until it produces a large mass of cells that we call a cancer. There is evidence that at least *two* mutations are required in the same cell to produce a cancer, which may be why it takes many years of exposure to low doses of radiation before a cancer develops. We thus have the remarkable situation that *whereas very high doses of ionizing radiation will cure cancer—as in radiation therapy—low doses over a long period of time can actually induce cancer.*

Besides somatic cells, our bodies have **germ cells** (eggs, sperm, and the cells that produce eggs and sperm). If one of these germ cells undergoes mutation and is then involved in the production of a child, the damage will be *inherited* and will occur in future generations. Such mutation may produce a physical malformation. But it may not have any visible effect at all, producing instead perhaps a lowered intelligence, or a genetic deficiency that leads to diseases like sickle-cell anemia or heritable cancer. An important distinction between somatic-cell and germ-cell damage is that the former affects only the person irradiated, while the latter affects many of the descendents of that person. Therefore, germ-cell effects are of far greater consequence for the population as a whole.

In summary, ionizing radiations damage the DNA and therefore the genes of cells. As a result, such radiation may

1. kill cells,
2. induce cancer, or
3. produce permanent deleterious changes in future generations.

This is why such radiation is dangerous.

Environmental Radiation: Effects of Low Doses

Our Earth, which seems to be so benign and stable on the surface, is in fact a seething cauldron of fluid rock and metal. We become aware of this only when a volcano erupts. Our seemingly stable continents are in fact rather thin plates of matter that have solidified because of heat loss to outer space, and they float about on the plastic rock beneath like slices of toast on the surface of a pot of thick melted cheese. If you go down 1 mile in a diamond mine, the temperature rises to 120° F—so hot that you need air conditioning to live. Dig down 20 miles into a continent, or 5 miles into the ocean floor, and you come to rock that is either plastic or molten. And if you go down all the way to the center of the Earth, the temperature will reach 12,000° F!

What keeps the Earth so hot? We used to think that it was residual heat from the days when the primitive Earth was entirely molten. But we now know that such primordial heat would have been radiated to outer space within a billion years. The Earth is over 4 billion years old. So where does the heat come from?

It comes from radioactivity. Many of the rocks and metals of the Earth are radioactive. Their atoms occasionally undergo nuclear rearrangements that give off powerful radiations. When these radiations are absorbed by matter in other rocks, they produce heat. Thus, the entire Earth is radioactive. *Radioactivity is a natural part of our environment, and it surrounds us all the time.*

We know that the plants we eat obtain many of their atoms from the Earth. And all the animals that we eat have eaten plants somewhere down the food chain. Furthermore, we drink water containing minerals that come from the Earth. So shouldn't we also be radioactive? Indeed we are! *We too are radioactive, like all living things on Earth.* It is estimated that the natural radioactivity in our bodies produces about 1000 cancer deaths per year in the U.S. This of course is a tiny fraction of the population, involving only about 1 person in 250,000.

As we have noted, "radiation" has become a horror word in today's world, partly because it is invisible and cannot be felt—and so seems mysterious—and partly because it indeed can be dangerous under certain circumstances. Most people, however, are unaware that radiation is a part of our natural environment.

The fact that biological organisms have lived with low levels of radiation throughout evolution suggests that such low levels are not very harmful. We must look into this matter more carefully. But first, we need to know how ionizing radiation is measured. The rather complicated nomenclature is described in Appendix B. However, we can get along most of the time by simply using the unit of **dose** called a **rem**. The rem is a measure of how effective the radiation is in producing biological damage. For example, some radiations, like alpha particles, produce a very dense track of ionization as they pass through tissue, whereas others, like beta particles (which are electrons), produce a very sparse track of ionization. The dense track is more effective in damaging DNA, so a single alpha particle traversing a cell is more biologically effective than a single beta particle.

How much radiation is represented by 1 rem? We can get some idea by noting that, if people are irradiated with 450 rems, over their whole bodies and in a short period of time, then 50% of them will die in 60 days. We say that 450 rems is the "lethal dose for humans." The dose response is sharp: usually, 300 rems will kill no one, but 600 rems will kill everyone.

Such *whole-body* irradiation is our primary concern in this book, which is about the effects of radiations from nuclear power plants and nuclear bombs. Much higher doses can be tolerated by small regions of the body, as in cancer therapy, where as much as 10,000 rems may be delivered to the tumor—which destroys it.

Natural Radiation Exposure

The radiation we encounter in our environment is what we call **natural background radiation**. It has four different sources (Figure 9):

1. Cosmic rays—charged particles from outer space that produce electrons and heavy electrons (**muons**) in our atmosphere.
2. Terrestrial radioisotopes—gamma rays from uranium,

thorium, and their decay products, in rocks and building materials, such as brick, stone, and sheetrock.
3. Internal radioisotopes—alpha, beta, and gamma rays, chiefly from potassium-40 and radioisotopes of polonium and lead, in our food, air, and water.
4. Radon—alpha particles from radon-daughter isotopes, which we breathe in as particles; the radon comes from radium in rocks and building materials.

Each of the first three contributes about $1/30$ rem per year, so that together they amount to about $1/10$ rem per year, but radon alone contributes another $2/10$ rem per year. Thus, *natural background radiation results in the exposure of Americans to about $3/10$ rem per year*, or about 21 rems in a 70-year lifetime.

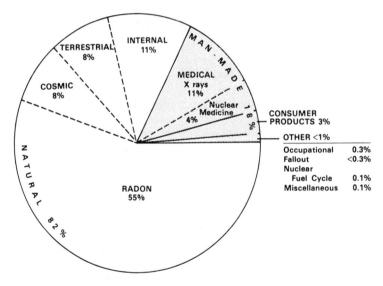

Figure 9. The percentage contribution of various radiation sources to the total average effective dose equivalent[4] in the U.S. population. (NCRP Report 93, September 1987—see Bibliography. Reprinted by permission of the National Council on Radiation Protection and Measurements.)

These figures vary. Cosmic rays increase with altitude, so that citizens of mile-high Denver get twice as much cosmic radiation as people living at sea level. Airline crews may receive $1/2$ rem per year from cosmic rays.[1] Radiation from internal radioisotopes is fairly constant. The amount of terrestrial radiation and radon that one is exposed to depends on the nature and location of one's housing. A stone house can contribute three times as much radiation as a wooden house. In some areas, surface rock is unusually radioactive: the geological area called the *Reading Prong*, which stretches from Reading, Pennsylvania to north of Peekskill, New York (Figure 10), may produce doses to the lungs as high as 100 rems per year. This is not as bad as it sounds, for this is not whole-body irradiation. However, it is

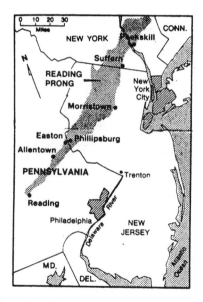

Figure 10. The uranium-rich Reading Prong, which emits high levels of radioactive radon gas. *(New York Times*, 28 October 1985. Copyright 1985 by The New York Times Company. Reprinted by permission.)

estimated that radon causes 5000–20,000 lung cancer deaths per year in the U.S.,[2] accounting for about 8% of all U.S. lung cancer deaths.

The latest analysis of survivors of the Japanese atomic bombings shows that the dose that will double the natural mutation rate in humans is roughly 200 rems,[3] or 10 times what a person gets from lifelong exposure to background radiation.

Imposed Radiation Exposure

In addition to natural background radiation, the average American is exposed to about 0.05 rem per year of medical radiation, much of it from fluoroscopic exams. One chest X ray gives only about 0.006 rem.[4] About 0.01 rem comes from consumer products, such as the radioactive sensors in smoke alarms. Figure 9 shows that, for commercial nuclear power, *the entire nuclear fuel cycle, from mining through power production and waste disposal, produces only 0.1% of the natural background radiation dose.* These data, from the National Council on Radiation Protection and Measurements (**NCRP**) are based on actual radiation monitoring. Furthermore, the exposure of the U.S. general public to radiation from commercial nuclear power accidents has been negligible.

Adding both natural and imposed radiation, we see that the average American receives a *total* of about 0.36 rem per year, or about 25 rems in a 70-year lifetime.

Effects of Low-Level Exposure

How much radiation can we take? This is a very important question, relevant to many of the issues discussed in this book. But it is not easy to answer.

Most of our information about the biological effects of radiation comes from data on people or animals that have been exposed to high doses. The reason is simply that it is difficult to measure radiation effects in humans at doses below about 50 rems. It is possible to detect human exposure to only a few rems by observing *chromosome aberrations* (changes visible under the microscope) in the white blood cells, but such detection requires comparison with *prior* measurements. Furthermore, in many cases, the lower the dose, the longer the radiation effect takes to show up. So, if a person receives a low dose at age 30 and comes down with cancer at age 60, how do we know that the cancer was radiation-induced? Or can we decide that he would normally have got cancer at age 62 but got it at age 60 because of his exposure to radiation?

Answers to such questions are almost impossible to obtain. Therefore, scientists study the effects of high doses of radiation and extrapolate them to low doses, assuming that radiation effects increase with dose in a linear manner. This simply means that, if 100 rems produces a given effect, then we can assume that 1 rem will produce $1/100$ of that effect. This process is known as *linear extrapolation to zero dose*, and it implies that there is no dose so low that there will not be at least some, although tiny, effect. An alternative to this **linear hypothesis** is the idea that radiation effects may be smaller than expected at low doses: a *sublinear effect*—or that there may even be no effect at all if the dose is sufficiently low: a *threshold effect* (Figure 11).

A recent study by the U.S. National Academy of Sciences concludes that the induction of solid tumors (tumors outside the circulatory system) fits the linear hypothesis, whereas the induction of leukemia (a cancer of the white blood cells) shows a sublinear effect.[5] On the other hand, among atomic-bomb survivors, threshold effects have been reported for several cancers, and deaths due to solid tumors have been surprisingly low.

How can we account for such lower-than-expected effects at low doses? For one thing, we know that many cells can repair much of the radiation damage to their DNA by using special enzymes, provided that the damage is not too great. Both damage that can kill the cell (lethal damage) and damage that can cause cancer (mutational damage) can be repaired. In retrospect, the existence of such repair systems is not so surprising.

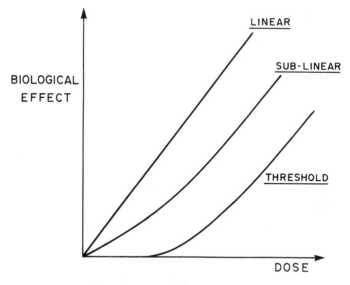

Figure 11. Diagram of a linear, a sublinear, and a threshold response to radiation.

We know that radiation levels must have been much higher billions of years ago, because the radioisotopes have all been decaying ever since the formation of the Earth. Yet life evolved in this higher radiation environment. One can only conclude that living organisms must have discovered some way of dealing with radiation damage.

One may conclude that *low doses are often less harmful, per unit dose, than high doses*. Put another way, we can say that many radiation effects show a sublinear dose response, and a few may even show a dose threshold, below which there is no effect. One caution should be noted: This conclusion is probably true for X rays and gamma rays but is probably not true for densely ionizing radiations, such as neutrons and alpha particles. Alpha particles, you will remember, are the radiations we get from exposure to radon in our houses.

It is of great interest that mutational damage leading to abnormalities has not shown up in the children of the atomic-bomb survivors or in the children of mothers who were in the Chernobyl area at the time of the accident. Such effects would be expected to appear during infancy. This outcome suggests that radiation damage to the germ cells has been repaired in these victims.

In summary, although the low doses we get from environmental radiation are, in principle, deleterious, our bodies appear to have learned to cope with much of the damage, just as we cope with poisons and twisted ankles. Of course, too much poison, or ankles twisted too often, may lead to permanent damage, such as blindness or chronically painful joints. So, too, excessive radiation is unquestionably bad for us, but low doses over long periods of time seem to be fairly innocuous.

Radiation Exposure Standards

How much radiation exposure is reasonable? It must be recognized that any answer really is a matter of drawing an arbitrary line. We are all exposed to natural radiation in the environment, and the amount may vary considerably from place to place. So it becomes an arbitrary thing to decide how much additional radiation is all right and how much is not. Fortunately, we do have some guidelines.

One might limit *additional* individual exposure to an amount equal to the average natural background level of radiation. Many people get quite a bit more than the average background simply by living at high altitudes, or in stone houses, or on rocky ground containing uranium or radium (which release both gamma rays and radon gas). Many residents of Colorado are exposed to all three of these elevated levels of radiation, yet there is no evidence that citizens of Colorado have higher cancer rates than citizens of Florida.[5] Swedes get *twice* as much background radiation as Americans, chiefly from higher radon levels, but do not have higher cancer rates than Americans.

For *the general public*, the NCRP has established a limit for additional *infrequent* radiation exposure, other than medical, of $1/2$ rem per year, which is about twice the background rate.[6] The National Academy of Sciences estimates that, if a population of 100,000 people were exposed to such a dose of $1/2$ rem of additional radiation per year, 40 of them would eventually die of radiation-induced cancer.[5] This is one person in every 2500, and represents only 0.2% of the 21,000 people who would normally die of cancer in this population. A level of $1/2$ rem per year therefore seems a not unreasonable exposure limit.

For *radiation workers*, the NCRP has established an absolute upper limit of 5 rems per year, including internal and external irradiation. This is 17 times the natural background. However, other limits are also set. For example, the accumulated lifetime dose is not to exceed the age of the person in rems—i.e., 25 rems for a 25-year-old, 50 rems for a 50-year-old, etc.

Finally, we must be aware of the difference between damage to germ cells and damage to somatic cells. Somatic-cell damage can be devastating, but it will affect only the individual exposed. Germ-cell damage, on the other hand, may be transmitted to future generations. Clearly, one expects to have lower "permitted doses" for germ cells. Lower permitted doses would also be expected for embryos, in which just a few cells may determine an entire future structure of the child's body, such as the brain or the heart. Indeed, the NCRP has set much lower limits for pregnant women than those noted above. Finally, if one is dealing with whole-body irradiation that is imposed on a population, such as that from nuclear power technology, then lower limits are also in order.

Much of our extra radiation exposure is medical, and about 50% of this involves some exposure of the gonads[7]—the ovaries and testicles that produce the germ cells. Even though the genital area may not be directly irradiated, radiation can be "scattered" by the X-ray machine or by body tissues, so that it is deflected into the gonads. However, during medical irradiation, exposure of the gonads can be greatly reduced by irradiating as small an area as possible, and by shielding the abdomen with lead aprons and the like. It should be noted that shielding of the gonads during the medical exposure of young people is much more important than for people who are beyond child-producing age. Abdominal fluoroscopies may involve quite high doses, but they are usually done on elderly people for such things as diagnosis of cancer. Although such exposure carries some chance of inducing cancer in the individual, it is of no long-term relevance for the population as a whole.

In this chapter, we have shown that radiation is a normal part of our environment, and one that living organisms appear to have learned to deal with. We have learned how ionizing radiation damages the DNA of living cells and may thus kill an organism or induce cancer. It may also cause heritable damage in germ tissues, but experience has shown this damage to be smaller than expected, probably because of DNA repair by enzymes within the cells. We now proceed to a discussion of how these actions of radiation are actually put to good use in medicine and agriculture. In addition, ionizing radiation has become a very useful tool in industry.

5 Radioisotopes in Medicine and Industry

In 1864, the Scotsman, James Clerk Maxwell, one of the greatest theoretical physicists of all time, proposed on purely theoretical grounds that light is an electromagnetic phenomenon. He predicted that there must exist a whole spectrum of electromagnetic radiations similar to light but having different energies. This bold prediction was supported by the discovery in Germany in 1887 of radio waves by Heinrich Hertz. Radio waves were found to penetrate matter easily, but had very low energy compared to visible light. In 1895, another German scientist, Wilhelm Roentgen, using an apparatus much like a modern television tube, discovered X rays. These rays also penetrated matter easily, but were far more energetic than visible light.

These revolutionary discoveries were followed in 1896 by a bigger surprise, when Henri Becquerel, at the University of Paris, discovered that uranium ore gives off radiations—later called radioactivity—far more powerful even than Roentgen's X rays. Two years later, his colleague Marie Curie isolated the first radioactive element from uranium ore; she named it *polonium*,[*]

---

[*]Not to be confused with *plutonium*, a transuranic atom (Appendix A).

after her native Poland. With her husband Pierre, she then went on to purify a second radioactive substance from the ore, which they called *radium*. They found that pure radium gave off rays *a million times as intense* as those of the uranium ore. It now began to dawn upon scientists that almost inconceivable energy exists inside atoms.

Applications followed quickly. Before the end of the century, a mobile X-ray unit was used by the British army in the Sudan to detect bullets and shrapnel in the wounds of soldiers. Radiation was first used to treat cancer by the removal of a skin tumor in 1899. Physicians in those days apparently did not recognize how damaging these radiations could be, and often exposed themselves needlessly to X rays. Many of them developed deformed hands or lost their fingers; some contracted bone cancer and leukemia. Both Marie Curie and her daughter, Irène, died late in life of leukemia, apparently induced by the radiations they had studied all their lives.

In 1934, Irène Curie and her husband, Frédéric Joliot, produced the first **artificial radioisotope**. By bombarding aluminum with alpha particles produced by radioactive polonium, they produced a *new radioactive atom* that had not before existed: phosphorus-30, with a half-life of 2.5 minutes. We now recognize that new isotopes are frequently produced when matter is bombarded by high-energy particles, and that many of these new isotopes may be radioactive. Thus, in a nuclear explosion, many atoms in structures on the ground, as well as in the dust raised from the ground into the mushroom cloud, become radioactive as a result of the tremendous numbers of neutrons released during the explosion. When a nuclear reactor has reached the end of its useful life of about 40 years, part of the waste-disposal problem is caused by the neutron-induced radioactivity in the steel components of the reactor core. On the other hand, artificial radioisotopes of a wide variety have become extremely useful in biological research and medicine. These isotopes are produced by neutrons from nuclear reactors, and by high-energy charged particles from particle accelerators—machines sometimes called "atom smashers."

By the year 1900, the scientific world had been bowled over by the new discoveries of nuclear energy. But more was to come. In 1905, Albert Einstein published his theory of special relativity, in which he predicted that *matter can be transformed into energy*. His calculations showed that the energy release would be beyond all previous experience of humankind. Physicists predicted that a few pounds of matter could drive the *Queen Mary* across the ocean. Many eminent scientists laughed at this idea. But in 1945, a chunk of uranium smaller than a basketball destroyed the city of Hiroshima. People stopped laughing.

In this book we are concerned about present and future applications of this vast energy resource. Few would argue that a nuclear war would be a vast catastrophe. On the other hand, the controlled release of nuclear energy to produce electricity has both desirable and undesirable aspects—with both of which we shall be concerned. But one area in which the use of nuclear energy is universally accepted as worthwhile and good is in biological research and in medicine and industry. These applications of nuclear energy are an unqualified success story. As we trace these applications, you will come to realize how deeply nuclear energy pervades our modern civilization.

Radioactive Tracers

In the first 40 years of this century, a new breed of chemists appeared, called biochemists, who studied the molecules of living systems and their chemical reactions. They determined the structures of amino acids and proteins, and learned how these molecules are built up and broken down by enzymes, as we have described in Chapter 2. Before 1940, these workers dealt with isolated and simple chemical systems. They learned a prodigious amount about pure biochemicals and how they react in test tubes. This research was very necessary groundwork, but it was a far cry from knowing how molecules behave inside complex living cells. Living cells are fragile and cannot be treated with strong chemicals. Their complexity and fragility made the goal of understanding what went on inside them seem almost unattainable. But gradually, after 1940, biochemists developed new techniques, each of which provided a little more information about living systems. These scientists then began to call themselves molecular biologists. One of their most powerful new techniques was the use of **radioactive tracers**.

Suppose you drink a liquid that contains some iodine atoms, such as a solution of the salt sodium iodide, which is not much different from sodium chloride, or table salt. In our bodies, iodine is used almost exclusively by the thyroid gland, which lies in the front of the neck. Most of the iodine you drink will quickly accumulate in your thyroid. But how will we know it is there? A surgeon could remove some of the gland, and we could then detect the iodine in this tissue by chemical means. But to show that the iodine was located *only* in the thyroid, one would have to operate on all the other tissues of your body—clearly an impossible treatment. But suppose that just a *tiny fraction* of the atoms in the solution you drank were not the usual isotope of iodine (^{127}I), but the radioactive isotope ^{131}I. Remember that all isotopes of a chemical atom show exactly the same chemical behavior, because they all have the same electron structure—only the number of neutrons in the nucleus is different. So ^{131}I behaves chemically just like the normal nonradioactive isotope of iodine, and wherever the normal iodine goes, the radioiodine will also go. After waiting for the iodine to accumulate in your thyroid, one could simply place a radiation counter close to your neck and it would start clicking, showing that the radioiodine was there. Placing the counter near other parts of your body would not produce this clicking, showing that the iodine was *not* there. In this very simple and elegant way, involving no surgery at all, one could answer the question of where the iodine goes in your body.

Of course, certain things must happen before such an experiment will work. First of all, the radioisotope used must emit gamma rays, as the short range of the alpha and beta rays will not allow them to escape from the thyroid and penetrate the overlying tissue. [131]I gives off such gamma rays, as can be seen from Table 1.

Second, the isotope must have a short half-life (implying a high decay rate) so that measurements can be made from such a tiny amount of the radioisotope that the radiation damage to biological tissues will be negligible. Iodine-131 has such a short half-life: only 8 days.

Third, where can you get radioactive iodine? We get it from

Table 1. Some Radioactive Isotopes Important
in Biology and Medicine

	Isotope	Half-life	Emission(s)	Energy[a] (Mev)
[3]H	Tritium	12 years	Beta	0.02
[14]C	Carbon-14	5600 years	Beta	0.16
[32]P	Phosphorus-32	14 days	Beta	1.7
[35]S	Sulfur-35	87 days	Beta	0.17
[40]K	Potassium-40	1.3 billion years	Beta	1.3
			Gamma	1.4
[60]Co	Cobalt-60	5 years	Beta	0.3
			Gamma	1.3
[90]Sr	Strontium-90	25 years	Beta	0.6
[99m]Tc	Technetium-99m[b]	6 hours	Beta	0.14
			X ray	0.14
[125]I	Iodine-125	60 days	Beta	0.03
			X ray	0.04
[131]I	Iodine-131	8 days	Beta	0.6
			Gamma	0.4
[192]Ir	Iridium-192	74 days	Beta	0.7
			Gamma	0.6

[a]Maximum energies of the emitted particles or rays are given in millions of electron volts (Mev). For comparison, a photon of visible light has an energy of about 3 electron volts.
[b]Technetium-99m is an unusual "metastable" (m) isotope. Both it and ^{125}I emit primarily X rays, which have lower energy than gamma rays.

nuclear reactors, the same kind of reactors we use to produce electric power. Some radioisotopes, like ^{131}I, are normal products of the fission of uranium, and the ^{131}I is obtained from used reactor fuel. Other radioisotopes can be made by bombarding ordinary atoms with the abundant neutrons from nuclear reactors. We discussed earlier the Joliot-Curies, who got radioactive phosphorus-30 by bombarding aluminum with alpha particles. Some radioisotopes are made by particles produced in particle accelerators.

Biological Research

Scientists can learn a great deal about *living* systems, using these "radioactive tracers," so called because they "trace out" the route taken by a particular atom. Furthermore, this technique requires such a tiny amount of radioisotope that it is usually quite harmless to the living system being studied.

Tracing where an atom goes is interesting enough, but one can also trace where molecules go. It is not hard for a chemist to make some protein precursor molecules, like amino acids, that contain a few radioactive sulfur-35 atoms, or some DNA precursors, like nucleotides, that contain a few radioactive phosphorus-32 atoms. If these are fed to a living cell, one can see where the radioactivity goes, and thus where protein is synthesized in the cell and where DNA is synthesized (protein contains sulfur but no phosphorus, and DNA contains phosphorus but no sulfur). In fact, this is precisely how molecular biologists learned that DNA is made in the cell nucleus, and that protein is made mostly in the cell cytoplasm.

In modern molecular biology, this tracer technique has probably yielded more new knowledge than any other technique. This knowledge gives us a better understanding of human health and the treatment of disease, an area called **nuclear medicine**.

Nuclear Medicine

Therapy

We discussed above how radioactive iodine could be used as a tracer to show that iodine is sequestered by the thyroid gland. The amount of radioactivity in such an experiment is too small to have any measurable effect on the thyroid. If a patient has a thyroid tumor, on the other hand, it can be treated by administering a *massive* dose of iodine-131. The radioisotope will quickly concentrate in the thyroid and, before it decays (its half-life is 8 days), it will heavily irradiate the thyroid tumor. Both the beta rays and the gamma rays will participate. The radioiodine will destroy not only the tumor but also any *metastases*, or subsidiary growths, in other parts of the body, because these metastases are made of thyroid tissue and will also take up the radioiodine. Thyroid tumors are routinely and successfully treated in this form of **radiation therapy**. In such therapy, the thyroid typically receives a dose of 10,000 rems or more. This dose may seem very high, but one must remember that the 450-rem lethal dose for a human (Chapter 3) involves total body exposure; much higher doses can be tolerated if administered to a small region of the body. That small region may, of course, be destroyed in the process, which is precisely what happens in cancer therapy.

Most radiation therapy, however, employs *external* sources of radiation, such as X-ray machines, with which we are all familiar. Another very useful machine uses the radioisotope cobalt-60. Table 1 shows that this emits a powerful gamma ray, with an energy of 1.3 million electron volts (Mev)—about a million times more energetic than a photon of visible light. Figure 12 shows a cobalt-60 unit. The highly shielded cobalt is located in the "head" of the device; when a shutter—made of lead—is opened by remote control, the radiation is directed downward at the tumor of a patient lying on the table.

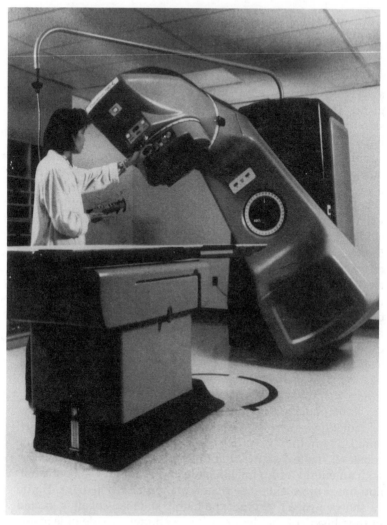

Figure 12. Cobalt-60 unit used for the radiation therapy of cancer. The nurse leaves the room before the radiation source is exposed, and she views the patient through thick "lead glass" windows, which shield her from all radiation. (Courtesy of Theratronics International, Ltd., Kanata, Ontario, Canada.)

One problem that arises in the radiation therapy of cancer is the irradiation of normal tissue. For example, the gamma rays from iodine-131 have the considerable energy of 0.4 Mev (Table 1). These penetrating rays will go beyond the thyroid and irradiate normal tissues outside the thyroid gland. But this results in a dose of only about 14 rems to the rest of the body.[1] The problems are much greater with an external radiation source, where the beam often has to travel through normal tissue before reaching the tumor. There are two ways to minimize this exposure of normal tissue. One is to use a source that rotates around the patient's body, but that is always aimed at the tumor. In such *rotation therapy*, the tumor receives a high dose, while any one region of the overlying normal tissue gets a much lower dose. A second way is to use special sources of radiation, such as cobalt-60, whose very energetic gamma ray penetrates deeper into human tissue than X rays. One may also use a source of very energetic particles, such as electrons, neutrons, or protons. Such particles tend to lose more of their energy near the ends of their ranges, thus lowering the "skin dose" and raising the "tumor dose." Machines producing these particles (such as cyclotrons, betatrons, and linear accelerators) are now used in radiation therapy, but they are expensive and are found only in large medical centers.

You may be wondering how it is possible that radiation in low doses may produce cancer, whereas in high doses it can cure cancer. This phenomenon is really not so strange. If you constantly walk barefoot on sharp stones, your feet will develop calluses. But with a very sharp stone, you could cut the callus off. Low doses of radiation alter the molecular structure of DNA; this may cause the irradiated cells to grow uncontrollably, that is, to produce a cancer. High doses of radiation produce so much damage in the DNA that they kill cells and may thus destroy a cancer.

In addition to *radiation therapy*, cancers may be destroyed by other agents that kill cells, such as chemicals—whence the term *chemotherapy*—or the tumor may be removed by *surgery*. Usually, one of these techniques alone does not eliminate the cancer, so tumors are often treated with a combination of two, or even all three, techniques. In treating breast cancer, for example, the breast and the adjoining lymph glands are often removed surgically. This surgery may be followed by X-ray treatment to kill any tumor tissue that the surgeon may have missed. If it is suspected that the tumor has started to spread to other parts of the body, then chemotherapy may also be used.

All of these cancer treatments involve debilitating side effects. Radiation therapy and chemotherapy may induce extreme nausea and may cause a loss of hair. We are all familiar with the side effects of surgery, many of which result from anesthesia, which fortunately does not have to be administered with radiation therapy or chemotherapy. Many patients feel that these side effects, drastic as they may be, are preferable to death by the cancer. But some patients choose not to have such therapies, since life may be prolonged for only a few months or perhaps a year, and the side effects lower the quality of life. In some cases, however, such therapies are highly successful. Cancers near the surfaces of body tissues, such as skin cancer and mouth cancer, are highly curable by radiation therapy, whereas deep cancers, such as those of the stomach and pancreas, are not.

Although most radiation therapy involves external sources, internal sources may be used, as noted above in the use of radioiodine to treat thyroid cancer. Radiologists often *implant* radioactive sources directly into the tumor. This technique has the virtue of putting the radiation source right where the cancer is. Wires of iridium-192 (Table 1) are now widely used because they are flexible and easily inserted. These sources must later be removed because the strong gamma rays that are emitted would harm the patient's normal tissues and perhaps even other people who come very close to the patient. "Seeds" of iodine-125—tiny tubes containing the radioisotope—are sometimes used for prostate tumors. They may be left in the patient permanently, because the radiations are very weak and will be gone after a year. Implants are usually used for tumors that are "accessible" but not on the outer surface of the body, such as those of the cervix or the rectum.

Diagnosis

Radioisotopes are also used to diagnose disease. We have noted the use of radioiodine to localize the region in which the iodine accumulates. Similarly, one may feed a patient certain molecules containing radioactive technetium-99m, which will localize in a tumor that is growing rapidly. A *gamma camera* uses radiation counters to record the radiation from a tumor that has been treated with technetium, giving a hazy, but extremely informative, picture of the tumor or other site. Brain and bone cancers are nicely diagnosed in this way. When used to detect cancer in bone, this technique is called a *bone scan*. Because of the 6-hour half-life of technetium-99m, the radioactivity quickly disappears.

Radioisotope techniques are now an indispensable tool in the diagnosis of disease. They are used in a variety of often complex diagnostic procedures, such as radioimmunoassay and positron emission tomography, discussion of which is beyond the scope of this book.

Industrial Applications

There are many nuclear radiation applications in fields outside cell biology and medicine. *These pose no hazard or risk to the public, and usually no risk to the user.* They are truly peaceful uses of nuclear energy.

Many of these applications involve only the use of X-ray machines. Others require more penetrating radiation, in the form of gamma rays, which are usually produced by a device called a *cobalt bomb*. Contrary to the suggestion in its name, this device does not explode! It is simply a mass of pure cobalt-60, which emits energetic gamma rays. It is enclosed in a thick lead container, to shield users from the radiation, but it has a "port" that can be opened by remote control, permitting the radiation to emerge in a specific direction. It is widely used in radiation therapy, as shown in Figure 12, and it also has many industrial applications. For example, cobalt-60 units are used in oil drilling to detect invisible cracks in drills and pipes.

Some industrial applications require neutrons. Neutrons may be obtained from a nuclear reactor, but this of course requires that the procedure be conducted at the site of the reactor. When a portable source of radiation is needed, one may use a small quantity, perhaps the size of a thimble, of californium-252. This isotope is one of the transuranic elements noted in Chapter 3, which are man-made elements of a higher atomic number than uranium. Californium-252 undergoes spontaneous fission, emitting copious amounts of energetic neutrons and gamma rays. As with the cobalt bomb, massive shielding is required. But, as in the cobalt bomb, the half-life is relatively short—2.6 years—so that the problem of disposing of old equipment is minimized.

Sources like californium-252 may also be used in a technique called **neutron activation analysis**, which uses neutrons to induce artificial radioactivity in a sample, thus revealing substances present in only trace amounts. After a sample has been activated (made radioactive) by the neutrons, one measures the radiation emitted by the now-radioactive sample. Every radioisotope has its own "signature" of radiation types, energies, and decay rates, permitting one to identify a chemical at extremely low concentrations. The level of radioactivity induced is very low and poses no biological hazard.

Geology

Neutron activation analysis is used in *mineral exploration*. A source of californium-252 is lowered into a bore hole. Above the source is extensive shielding (lead for the gamma rays, and paraffin for the neutrons), and above that is a radiation detector (Figure 13). As the californium source is lowered, it induces radioactivity at some point in the adjacent rocks. When the radiation detector reaches that point, it measures this radioactivity, thus revealing the nature of the atoms present at that point in the rock. In this manner, an entire profile of the mineral content in the borehole may be obtained.

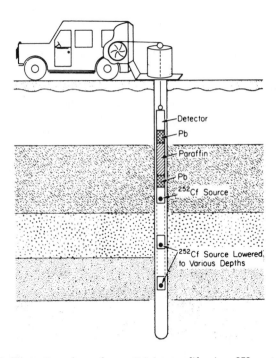

Figure 13. Illustration of a probe, containing a californium-252 neutron source and a radiation detector, being lowered into a drill hole for neutron activation analysis of the surrounding strata. (Reprinted with permission from Eric J. Hall, *Radiation and Life*, 2nd ed.—see Bibliography. Copyright 1984 by Pergamon Press PLC.)

Agriculture

Plants, like all organisms, undergo natural mutations from time to time. For millennia, people have selected or bred new varieties of plants by encouraging the growth of such mutants. Irradiation of seeds will tremendously speed up the mutation rate, resulting in *new plant varieties*. Most mutations produced by irradiation are deleterious, but occasionally, perhaps once in a million times, a beneficial mutation occurs. If millions of seeds are irradiated by X rays or gamma rays, an occasional plant derived from them will be found to have a desirable characteristic, such as growing higher or producing more fruit. Mutations produced in seeds are inherited, so seeds from a mutant plant will produce offspring that retain the desirable characteristic. In this manner, we have, for example, developed many new varieties of cereal grains, which have been critically important for the agriculture of underdeveloped countries.

Some insects have been successfully controlled by *insect sterilization*. Male flies are irradiated in the laboratory with gamma rays until they are sterile. When released into the environment, they mate with fertile females that mate only once, so these females produce no offspring. The release of sterile flies can result in the total elimination of an insect pest population. The technique has been successfully applied in the Gulf states to limit the screwworm fly, whose larvae infest the wounds of domestic animals. It has met with less success in attempts to control the Mediterranean fruit fly in California. The advantages of this sterilization approach over the use of chemicals are numerous: (1) no poisons are put into the environment; (2) only the target insect is destroyed, thus saving other insect species which may constitute the food supply of valued insect predators, such as birds; and (3) the technique can be readily applied in underdeveloped nations, which may be unable to afford expensive insecticidal chemicals.

Sterilization of food is another application of nuclear radiations. Much of the harvest of crops, especially in underdeveloped nations, is lost to spoilage due to insects, bacteria, and molds. Irradiation with gamma rays can prevent this spoilage, and is quite feasible, even though the doses required are high. No deleterious effects have been observed. Fruits are particularly suited to this technique, which enables many tropical varieties to be shipped to northern countries. The food can also be sterilized by irradiation at later stages, even after the food has been packaged, as the gamma rays will easily penetrate the packaging. Radiation sterilization of food for personnel of nuclear submarines has been practiced in the United States for many years. Sterilization of food with gamma radiation is likely to increase in the future; it is harmless and superior to the use of chemicals, which are often harmful.

It is important to recognize that such sterilization is done with radiations, like gamma rays, that produce ionizations in, and thus kill, bacteria and insects present in the food, but that cannot induce radioactivity. The induction of radioactivity requires particles, like neutrons or protons, that can penetrate the atomic nucleus. Consequently, *there is no radiation hazard whatsoever to the consumer of food that has been irradiated with gamma rays—such food is not radioactive.* However, the high doses used may alter chemicals in the food and may possibly change the taste. This seldom happens, but it occurs notably in sterilization of milk. This taste change can be avoided by irradiating the milk with ultraviolet light instead of with gamma rays. It is suspected, but not yet proved, that irradiation of food may destroy some nutrients, like vitamins. However, such a loss would certainly be less than is produced by the heating required for canning.

Other Areas

A wide variety of other applications exist for nuclear radiation. Cobalt gamma rays are used to detect flaws in metal parts during manufacture. They are also used for the sterilization of medical supplies, which may be done after packaging, thus killing any contamination produced by human contact.

Neutron activation analysis is used in archaeology, where the geographical source of an artifact can be reliably estimated. It is also used in forensic medicine, where, for example, a piece of clothing or hair found at the scene of a crime can be unequivocally matched to those of a suspect. Recently, arsenic detected by neutron activation analysis in preserved hairs of Napoleon led to the theory that he died of poisoning.

There are two great virtues of neutron activation analysis: only tiny samples are needed, and the technique does not damage the sample.

In this chapter, only some of the most widely used applications of nuclear radiations have been discussed. Yet even this brief survey reveals how pervasive these techniques have become in our modern civilization. As indicated above, this is a success story. Rarely are any hazards associated with these techniques.

Many people oppose nuclear war, and rightfully so—it presents a greater hazard than humankind has ever faced. And many people extend this horror of nuclear war to a horror of nuclear power. There is now a great debate on nuclear power in this country, a concern that we address in the next section of this book. But there can be no debate about the peaceful, extremely useful, and generally harmless applications of nuclear energy noted in this chapter. Those who oppose *all* use of nuclear energy in *all* walks of life are simply not being rational.

III ⚛ The Power

6 Nuclear Creation

In 1905, Albert Einstein predicted mathematically that matter could be converted into energy. The world has never been the same since.

What did he mean? An example of this phenomenon occurs routinely in high-energy physics laboratories. There is a kind of matter in the universe called *antimatter*. This is made up of atoms that have antiprotons, antineutrons, and antielectrons. An antielectron is called a *positron*; it behaves in every way like an electron, having the same almost-zero mass, but it has a *positive* electric charge. Now, if a positron and an electron collide with each other, they *annihilate* each other. They vanish! All of their matter, or mass, disappears. They are converted completely into two energetic gamma rays, which fly off in opposite directions. Gamma rays have no mass and are pure energy. These gamma rays are called *annihilation radiation*.

This was one of the most remarkable discoveries of modern science. It means that nothing is permanent. Classical science speaks of the Law of Conservation of Matter, meaning that matter is never lost in chemical reactions; it is simply converted into another form, as the gasoline that you burn in your car is converted into hot gases and carbon deposits. This conservation principle still holds in our everyday life. But nuclear reactions can *destroy matter*, thus contradicting the Law of Conservation of Matter. If our Earth were ever to encounter another Earth made of antimatter, the two would annihilate each other, and nothing would be left over but energy. Fortunately, there isn't much antimatter near us in the universe, although we create tiny amounts of it in the laboratory. It may exist in large quantities in some parts of the universe.

An equally astounding phenomenon involves the *creation of matter*. When a high-energy gamma ray passes through matter, it may "graze" an atomic nucleus. As it does this, it can create an electron and a positron, using some of its own energy—thus creating brand-new matter where none existed before. This phenomenon is the opposite of annihilation, and is called *pair production*. Pair production violates the classical Law of Conservation of Energy.

However, the "new physics" gives rise to a new conservation law: the *Law of Conservation of Matter-Energy*. That is, whenever some matter is destroyed, an entirely predictable amount of energy is created, and vice versa, when energy is transformed into matter. This is expressed in Einstein's famous equation

$$E = mc^2$$

where E is energy, m is mass, and c is the velocity of light.* Matter is converted into energy in **nuclear reactors** and in **nuclear bombs**. Because we can calculate how much mass is lost in such devices, Einstein's equation permits us to predict how much energy nuclear reactors and bombs will produce.

All nuclear reactions involve conversions of matter and energy. In radioactivity, the particles that fly out from the nucleus (electrons, protons, and alpha particles) are energetic, which is to say, they are moving fast. They get this energy from the conversion of a very small amount of matter within their nuclei. So, the first observation of the conversion of matter into energy

---

*The velocity of light is very great: 186,000 miles per second, or 670 million miles per hour. It is *constant* and can never be exceeded by any material object; this limitation places a constraint on space travel to distant stars.

predated Einstein, and actually occurred with the discovery of uranium radioactivity by Henri Becquerel in 1896.

Nuclear Fission

There is a good deal of radioactivity around us all the time—in the stones of the ground, in the air, and even in our own bodies. But the energy is released slowly, from radioactive atoms that are at relatively low concentration, so most of this radioactivity is not dangerous.

However, there is a way in which much more energy can be released all at one time—by the splitting of an atom into two parts. We call this process **nuclear fission**. It was first recognized as such in a laboratory experiment by Otto Hahn and Fritz Strassmann in Germany in 1939. The fact that this discovery occurred in Nazi Germany largely accounts for the feverish wartime pursuit of nuclear weapons by America, which exploded the first nuclear bomb in 1945, only six years later.

Nuclear bombs were at the time inaccurately called "atomic" bombs. A TNT bomb is really an atomic bomb, because its reactions are at the atomic level. But **fission bombs** are *nuclear* bombs, the reaction being at the nuclear level. Nevertheless, by habit we still call the World War II devices **atomic bombs**.

The heaviest natural atoms, like uranium, undergo **spontaneous fission**. They are slightly unstable atoms, and their nuclei occasionally split apart. Indeed, the reason why the periodic table of *natural* atoms terminates with uranium (see Appendix A) is that all heavier atoms—the transuranic elements—have decayed, either by fission or by radioactivity, within the lifetime of the Earth, so that none of them are left. The atoms known to us in the periodic table beyond uranium have all been created by man. The most important of these is plutonium (Pu), which is made in nuclear reactors, and is used for both reactor fuel and nuclear bombs. Isotopes that undergo spontaneous fission include thorium-232; uranium-234, 235, and 238; and plutonium-239 and 242.

A heavy atom can also undergo **induced fission**, a process in which the absorption of a neutron by the nucleus makes it so unstable that it splits apart. Figure 14 shows what happens in the induced nuclear fission of the uncommon isotope uranium-235 (more than 99% of natural uranium is the isotope U-238). This U-235 fission reaction is used in most nuclear reactors and was used in the atomic bomb that was dropped on Hiroshima.

In this fission reaction, a neutron collides with the nucleus of uranium-235, and it enters the nucleus. It can do this easily because, although the uranium nucleus has a very great positive electric charge from its 92 protons, the neutron has no charge, and it is therefore not repelled by this positive charge. We now have a nucleus of uranium-236. With the extra neutron, this nucleus is unstable, and it immediately splits apart in a violent reaction, creating two new, much smaller nuclei, called **fission products**. In the example shown, these fission products are the nuclei of xenon-140 and strontium-94. In addition, gamma rays are produced, and two neutrons are released. The violence of the reaction, which may be compared to firing a cocked gun, causes the released fission-product atoms and neutrons to have

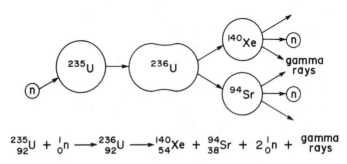

$$^{235}_{92}U + ^{1}_{0}n \longrightarrow ^{236}_{92}U \longrightarrow ^{140}_{54}Xe + ^{94}_{38}Sr + 2^{1}_{0}n + \text{gamma rays}$$

Figure 14. Diagram of induced nuclear fission of uranium-235.

very high energy, which means that they are moving very fast. In other words, a lot of **heat** is produced.

Heat is simply the random motion of molecules, atoms, or particles. In a hot gas, the molecules move very rapidly, and this in fact is what we mean when we say that the gas is "hot." These rapidly moving molecules will bounce off the walls of their container, which might be a balloon, and thus exert a **pressure** on the container walls. We all know that hot gas expands, and it does so because of the greater pressure. If we cool the gas, the molecules will move more slowly, the pressure will drop, and the balloon will shrink. In air at normal atmospheric pressure and temperature, the molecules of oxygen and nitrogen move with the speed of rifle bullets, but they move only a tiny distance before they hit another molecule. In a liquid, the atoms or molecules are closer together and slightly bound to each other, so they move around less—like a person trying to move in a crowd. In a solid, the atoms are so crowded that they do not move around at all, but simply vibrate—this vibration is a measure of the temperature of the solid. Regardless of the type of motion, *temperature is a measure of the degree of atomic or molecular motion*.

In a nuclear bomb explosion, the extremely fast-moving atoms and neutrons produced by uranium fission (arrows at the right in Figure 14) strike the molecules of the air around them, causing them in turn to move very fast or, in other words, to "heat up the air." This process accounts for most of the tremendous heat present in the fireball of a nuclear bomb explosion. The rapid expansion of the air caused by this heat produces a wave of air pressure called a *blast wave*. The heat and the blast cause most of the damage to buildings, people, or anything else near the explosion. However, considerable additional heat is produced by the gamma rays and neutrons that fly out from the fireball and are absorbed directly by the atoms and molecules of people and buildings on the ground.

How do we get a nuclear explosion? Note that, for every neutron that went into the uranium induced-fission reaction, *two* came out. If each of these two neutrons then enters another uranium nucleus, they will produce *two fissions*. These will be followed by 4 fissions, from the 4 neutrons now available, then 8, then 16, and so on. In short, we have a **chain reaction**. Rapid chain reactions produce explosions. A tremendous amount of energy is released in a nuclear bomb explosion, but the entire chain reaction takes less than a millionth of a second.

Such a chain reaction can, however, be limited, as in a nuclear reactor, so that heat is released in a controlled way. As we shall see later, nuclear reactors are so designed that they cannot explode like a nuclear bomb.

In the equation shown in Figure 14, the sum of the atomic weights (the superscripts) to the left of each arrow equals the sum to the right of the arrow, and the same holds for the atomic numbers (subscripts). This balancing is required in any nuclear equation, and is an expression of the Law of Conservation of Mass. Thus, the two daughter nuclei, xenon-140 and strontium-94, plus the two neutrons, have the same atomic weight as the parent nucleus, uranium-236 ($140 + 94 + 2 = 236$). (The "atomic number" of a neutron is zero and the "atomic weight" is 1.) But where is the lost mass that was converted into energy? It is, in fact, only a *fraction* of a single atomic weight unit, and therefore it does not show up in the rounded-off numbers used in the equation. This tiny amount of mass produces the tremendous energy released in nuclear fission.

The daughter nuclei, or fission products, are shown in Figure 14 as being xenon-140 and strontium-94. These are only typical fission products; the reaction may produce other nuclei, with atomic weights varying about ±10 units around 140 and 94, but their *sum* (234) will always be the same. Thus, we might have lanthanum-139 and molybdenum-95 ($139 + 95 = 234$). We cannot predict which particular pair of products will be produced in any one fission event. We might have iodine-131 and rhodium-103. This latter reaction, occurring in nuclear reactors, supplies us with the iodine-131 that we use in nuclear medicine.

Appendix C shows some typical fission products from nuclear reactors.

One may wonder why we do not use the common isotope of uranium, U-238, in reactors and in bombs. The answer is that U-238 will not support a chain reaction, primarily because it "captures" neutrons and is thus converted into plutonium-239 (Chapter 7, Breeder Reactors). Plutonium-239 is actually preferred to U-235 for modern bombs, because it releases more neutrons per fission, thus increasing the efficiency of the chain reaction.[1] Plutonium-239 was used in the Nagasaki bomb. It can also be used in nuclear reactors.

It is important not to confuse radioactivity and nuclear fission. The former is indeed a nuclear reaction, but it involves a small adjustment in the nucleus, releasing an alpha particle or a beta particle, as well as a relatively small amount of energy. Furthermore, it occurs slowly, one atom at a time. The daughter atom is very similar in weight to the parent atom. Nuclear fission, on the other hand, involves a violent splitting of the nucleus, releasing a tremendous amount of energy, and it can be made to occur explosively, in a chain reaction. The daughter atoms are much smaller than the parent atom.

Nuclear Fusion

An entirely different process from nuclear fission, but one that can be equally devastating, is **nuclear fusion**, illustrated in Figure 15. In *fission*, we break down a very heavy atom, near the end of the periodic table, into two smaller atoms. In *fusion*, we build up, or create, a new atom by fusing together two very light ones, usually hydrogen, the first atom in the periodic table.

Hydrogen fusion requires an extremely high temperature, roughly 100 million degrees Celsius (centigrade). In the reaction, two hydrogen nuclei fuse to produce a helium nucleus. Like nuclear fission, this process involves a *loss of mass*, with a conse-

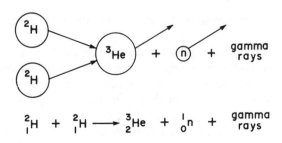

$$^2_1H + ^2_1H \longrightarrow ^3_2He + ^1_0n + \text{gamma rays}$$

Figure 15. Diagram of hydrogen fusion.

quent release of very great amounts of heat energy, in the form of high-energy neutrons and gamma rays. But unlike fission, fusion is not a chain reaction: the released neutron does not participate in a further reaction. Instead, the fusion process simply depends on a tremendous temperature to maintain the reaction. For this reason, fusion reactions are also called **thermonuclear reactions**. Nuclear fusion is a very difficult reaction to produce: it occurs only in stars, like our Sun, and in the **hydrogen bomb**, which is a **fusion bomb**. The only way we can get the high temperature needed to ignite a hydrogen bomb is to explode a fission bomb inside the hydrogen bomb. The fission bomb thus serves as a trigger.

You may note something curious about the reaction shown in Figure 15: the hydrogen has an atomic weight of 2, instead of the usual 1. This is an isotope of hydrogen (H), called **deuterium** (D or ^2H). Its nucleus contains a proton *and a neutron*. Thus, two atoms of deuterium fuse to form an atom of helium: the light isotope helium-3 (^3He).

When deuterium combines with oxygen to form water, it produces what is called deuterium oxide, or **heavy water** (D_2O). You will remember that isotopes of an atom all behave chemically like one another, so heavy water behaves chemically almost exactly like ordinary or "light" water. About 0.02% of the hydrogen in natural water, including rain and ocean water, is deuterium. Another isotope of hydrogen can be used in fusion reactions, namely **tritium** (^{3}H), which has an atomic weight of 3 (1 proton and 2 neutrons). Tritium can be produced in nuclear reactors. It would be exceedingly difficult to make a hydrogen bomb from normal hydrogen; we must use deuterium or tritium. In fact, the common method is to use a reaction that involves both: a deuterium nucleus and a tritium nucleus combine to produce a nucleus of normal helium-4 (^{4}He) plus a neutron.[2]

Fusion reactions are the basis of the hydrogen bomb, and they also are the way in which the Sun obtains its energy. *The center of the Sun is like a great hydrogen bomb exploding all the time.* This nuclear fusion is made possible by the extreme temperature at the center of the Sun. The Sun has been doing this for close to 5 billion years and will continue to do so for at least as long into the future! The Sun, like other stars in the universe, is composed mostly of hydrogen, which is being slowly converted to helium through nuclear fusion.

We are now attempting to produce nuclear fusion in the laboratory, in an effort to develop a *nuclear fusion reactor*. This would be a much cleaner source of energy for producing electricity than a fission reactor, and the fuel supply—deuterium in the oceans—is limitless.

The release of the awesome power of nuclear energy is an event that will be remembered in our most distant future history. It is comparable to the discovery of fire, and eclipses the importance of the harnessing of falling water and steam for power. Humankind will never forget this event. Nor will it forget that, within six years of its discovery, it was used to destroy cities. The release of this tremendous energy provides us with a truly historic opportunity for good, as well as an historic opportunity for evil. The choice is ours.

7 Nuclear Power Reactors

The nuclear fission of uranium-235 described in the last chapter can be used to make either a nuclear bomb or a nuclear reactor. In this chapter, we describe the design and operating characteristics of nuclear power reactors, those reactors whose function is to make electricity. In subsequent chapters, we shall consider reactor accidents and problems of waste disposal.

Whether you are making a bomb or a reactor, a sufficient amount of uranium-235 must be assembled to support a chain reaction. Natural uranium ore is mostly uranium-238, less than 1% of it being U-235. In such ore, the concentration of the U-235 is too low to produce a chain reaction under most conditions. For nuclear power reactors, the uranium is therefore usually *enriched*—meaning that the concentration of U-235 is raised—to a level of about 3% U-235.

Suppose we start with a small mass of such enriched uranium, say, the size of a pea. As we have noted, *spontaneous fission* of uranium-238 will produce a few neutrons, some of which will react with atoms of U-235 to produce *induced fission*. However, neutron collisions with a uranium atom that result in nuclear fission are very rare, so most of the neutrons will escape from the mass (fly off into space) before they have time to react with a nucleus of U-235 to start a chain reaction. But if you start with a much larger mass of uranium, the neutrons will collide with more atoms and will be deflected back into the uranium mass; thus the chance increases that they will produce fission before they escape. Figure 16 illustrates this situation. With a sufficiently great mass of uranium, a chain reaction occurs, and we say that a **critical mass** has been achieved. This is a bit of a misnomer, because it is clearly not just the mass that is important, but also the *shape* of the mass. For example, a spherical critical mass would no longer support a chain reaction if it were flattened out into a sheet, because then most of the neutrons would escape from the mass. If one has a mass greater than the critical mass, it is called a *supercritical mass*; if the mass is less than the critical amount, it is a *subcritical mass*.

Now, if you were to keep slowly adding enriched uranium until you achieved a critical mass, there would be a sudden huge release of radiation and heat, which could easily kill you. How, then, can you build a nuclear bomb? One way is to have two subcritical masses—such as two hemispheres—separated from each other in the bomb. When you wish to have the bomb ex-

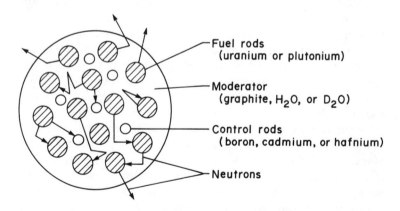

Figure 16. Cross-section of the reactor core of a nuclear reactor. Vertical (perpendicular-to-the-page) fuel rods and control rods are embedded in a neutron moderator.

plode, these parts can be driven together very rapidly, by means of a chemical explosive. When the two parts come together, the mass suddenly becomes supercritical, the chain reaction takes off, and the bomb explodes. The chain reaction is completed very rapidly, taking only a millionth of a second. The Hiroshima bomb worked this way. In any nuclear bomb, things are arranged so that the mass just before explosion is highly supercritical.

In a nuclear reactor, on the other hand, the reaction is controlled. This is done by limiting the number of neutrons available for fission, so that the reaction is just barely self-sustaining—the reactor is thus always in a **critical condition**, always on the verge of a chain reaction. A decrease in the number of neutrons available will cause the reactor to go subcritical, and it will slow down and stop. If the number of available neutrons increases rapidly, the reaction will go supercritical, and a chain reaction will develop: in some tens of seconds the reactor will go out of control, and the resulting great release of heat may result in a meltdown. As it melted, however, the reacting mass would change shape, quickly becoming subcritical, and the chain reaction would stop.

Nuclear Reactor Design

The high production of heat in nuclear chain reactions can be very useful, as most power plants work by producing heat, usually by burning coal, oil, or natural gas. In such power plants, the heat is used to produce steam, and the steam drives a turbine, which in turn spins an electric generator. The process is illustrated in Figure 17, where the heat is supplied by a nuclear reactor. In a nuclear reactor, the fuel consists of cylindrical pellets of uranium metal or uranium dioxide, about as big as the tip of your little finger. Hundreds of these are placed in a long vertical tube, called **cladding**, that is made typically of stainless steel or an alloy of the metal zirconium. About 200 of these **fuel**

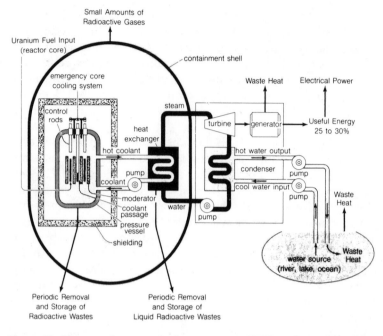

Small Amounts of
Radioactive Gases

Uranium Fuel Input
(reactor core)

containment shell

emergency core
cooling system

Waste Heat   Electrical Power

steam

control
rods

turbine   generator   Useful Energy
25 to 30%

heat
exchanger

hot coolant

hot water output

pump

condenser   pump

coolant

cool water input

moderator

pump

coolant
passage

water

Waste
Heat

pressure
vessel

pump

shielding

Waste
water source   Heat
(river, lake, ocean)

Periodic Removal
and Storage of
Radioactive Wastes

Periodic Removal
and Storage of
Liquid Radioactive Wastes

Figure 17. Diagram of a pressurized-water reactor (PWR) nuclear power plant. (From G. Tyler Miller, Jr., *Living in the Environment*, 2nd ed., copyright 1979 by Wadsworth, Inc. Reprinted by permission of the publisher.)

rods are grouped into an assembly called a *fuel element*. Some 180 of these fuel elements make up the cylindrical volume called the **reactor core**.

But how can we maintain a critical condition in a nuclear reactor without the whole thing melting down? This problem had to be solved before nuclear power could become feasible. Part of the answer is by using **control rods**. These are made of alloys containing the elements boron (B), cadmium (Cd), or hafnium (Hf). These metals have a very strong tendency to absorb neutrons and will therefore remove neutrons from the reacting mass. Neutrons that enter control rods are harmlessly absorbed but produce some heat in the process. By continually raising and lowering a few dozen of these rods in and out of the reactor core, one can maintain the reaction at any desired level. These control rods dampen the reaction without completely stopping it. You could compare it to spraying a mist of water on a barbecue fire.

It may have occurred to you that it would be necessary to move these control rods in and out very rapidly indeed, for a reactor, like a bomb, could go supercritical very quickly. However, referring to Figure 14, you can see that most of the neutrons are produced by the daughter nuclei (the fission products). The daughter nuclei release these neutrons as they decay to more stable isotopes. About 1% of this neutron emission takes a little while to occur; we say it is "delayed" for times ranging up to several minutes. Nuclear reactors are controlled—by the control rods—so as to just barely maintain a critical condition. In this situation, the continuing reaction *depends on the delayed neutrons*. This means that, should the control rods be withdrawn too far or too fast, one would have tens of seconds in which to plunge them back in again, before the fuel mass went out of control. Tens of seconds is a long time compared with a microsecond and is plenty of time to drop the control rods back into the fuel assembly to stop the chain reaction. Therefore, one of the important *inherent* safety factors in a nuclear reactor is its dependence on delayed neutrons.

Let us consider another aspect of the neutrons. The reaction of a neutron with uranium-235 occurs efficiently only if the neutron is moving with a speed similar to that of molecules at room temperature. Such neutrons are considered to be moving *very slowly* and are called **thermal neutrons** because their energy is similar to that of molecules at normal, or "thermal," temperatures. The neutrons produced by uranium fission are initially moving very fast, but if they are slowed down, they will split many more uranium-235 atoms. They can be slowed down by certain substances that remove much of the energy of motion of these neutrons (and are themselves heated up in the process). We call such substances neutron **moderators**. Graphite, which is pure carbon, and "light" water (H_2O), as well as "heavy" water (D_2O—see Chapter 6), are good moderators. Figure 16 shows a cross-section of a typical reactor core, in which the fuel rods and the control rods are embedded in a moderator.

Reactors that use U-235 require thermal neutrons. For this reason, and *not* because they produce heat (all reactors produce heat), they are called *thermal reactors*, or *slow reactors*. Some reactors that use other isotopes for fuel may be able to use fast neutrons and therefore do not need a moderator; they are called *fast reactors*.

Of great interest is a *natural reactor* recently discovered in the Gabon Republic of Africa by French engineers, who found a slight depletion of U-235 in uranium ore from a mine; they also found some fission products in the ore. Because U-235 is constantly decaying, its concentration in uranium ore about 1.7 billion years ago would have been much higher than at present (around 3%). Such ore could have occasionally gone critical and produced heat, just as a power reactor does, provided that water was flowing through it to act as a moderator. In times of drought, the "reactor" would have stopped. This natural reactor apparently produced power of several kilowatts, off-and-on, for at least 100,000 years! Because much of the U-235 has now decayed, both by radioactivity and by fission, the reactor no longer functions. It is estimated that there may be as many as 100 such natural sites in the world that once behaved as active nuclear reactors.

The function of a nuclear reactor is to produce heat, and one must have a liquid or gas—called a *coolant*—that will absorb this heat and transport it away from the reactor core. In most Western commercial reactors, including all American types, the moderator is water or heavy water, and *this water serves as both a moderator and a coolant*.

Figure 17 diagrams a pressurized-water reactor, the type most commonly used in the United States. In such a reactor, water—maintained at high pressure so that it will remain liquid even at very high temperature—removes heat from the reactor pressure vessel and is pumped away from the core. This hot coolant water then passes through the tubes of a *heat exchanger* (your car radiator is a heat exchanger), thus making those tubes very hot. Flowing around these hot tubes in the heat exchanger is a second circulating system of water at normal pressure, which is converted into steam by the very high temperature of the tubes. This steam turns a turbine, which, in turn, spins a generator, to produce electricity. The "used steam" that has passed through the turbine, and that has thereby lost much of its heat, is then condensed to water by passage through the tubes of a condenser—another heat exchanger—surrounded by the cold water of a *third* water-circulating system. This third water system is thus heated up, and it is then routed outdoors, where it is cooled either in *evaporating towers*—the big towers that one sees at most nuclear power plants—or by water from a river or a lake. It is then recycled into the system.

A heavy steel shell, or *pressure vessel*, about 9 inches thick, encloses the reactor core. In some designs, additional 8-foot-thick reinforced-concrete shielding surrounds the pressure vessel, the control-rod mechanism, and the emergency core-cooling system. All of this is in turn contained within a 3-foot-thick reinforced-concrete *containment shell* (usually a building), which may be lined with 4-inch-thick steel, that encloses all components of the pressurized-water system. The turbines, generators, and condensers are in a separate building.

About two-thirds of the nuclear power reactors used in the United States are **pressurized-water reactors (PWRs)**, and most of the others are **boiling-water reactors (BWRs)**. In a PWR, the water in the primary cooling system (the one that extracts heat from the reactor core) is maintained at very high pressure—about 140 times atmospheric pressure. Under such pressure, water will not boil even at the 600° F temperature that it acquires in the core, so the water stays liquid at all times. In a BWR, the pressure in the primary system is only about half as great, so that the water begins to boil as it passes through the core. A PWR needs three water systems: one that extracts heat from the core, one that receives this heat and boils to produce steam, and one that cools the steam system. In the BWR, the first system is not needed. The core-cooling system itself boils and produces steam; thus it behaves like the secondary water system in a PWR. In some respects, the PWR is a safer reactor, because production of steam inside the reactor core can be dangerous, as we shall see when we discuss the Chernobyl accident.

Safety under Normal Operations

Under normal operating conditions, nuclear power plants are very safe, partly because a great many *fail-safe* devices have been engineered into the design, and also because of certain *inherent* features of the reactor itself, especially the PWR:

1. *Reactor criticality depends on delayed neutrons.* This phenomenon, described above, provides extra time in which to lower the control rods if the nuclear reaction starts to go out of control.
2. *As the temperature of a reactor core rises, the fission process slows, and therefore the rate of heat production slows.* This occurs because (a) increased temperature causes the U-235 to expand, moving the uranium atoms farther apart, and making it less likely that a neutron will collide with a uranium atom, and (b) at higher temperature, U-238 absorbs more neutrons, thus draining neutrons away from the U-235 chain reaction. In addition, if the reactor cooling water also serves as the moderator, as in most Western reactors, then (c) as the temperature rises, the water molecules move farther apart, making it less likely that a neutron will hit them; the water is then a poorer moderator (i.e., less able to slow down the fast

neutrons), and this poorer moderation slows the chain reaction.

3. *In a PWR, circulating pressurized water does not leave the containment shell.* This water, which may become radioactive, is never passed through a turbine or condenser, as in the BWR, and therefore is less likely to contaminate other parts of the power plant.

What about the radiation and other toxic emissions from a nuclear power plant? In fact, they are almost nonexistent: neighboring populations typically receive less than 0.001 rem of radiation per year. *A normally operating nuclear power plant emits no air pollutants and typically less radiation than a coal-burning power plant.* People find this hard to believe, but it's true.[1] Let us look a little closer.

Most of the radioactivity produced in a nuclear power reactor is retained within the fuel rods. Tiny quantities of fission products, however, such as the noble gases xenon and krypton, may diffuse through the walls of the rods and enter the cooling water. In addition, the neutrons in the core produce some radioisotopes in the coolant system itself, such as tritium from the water and manganese-54 from neutron bombardment of the metal surfaces (Appendix C). Most of these radioactive products are filtered out of the cooling system; the liquids and gases are then held at the power plant for some weeks to let short-lived isotopes decay. Products released to the environment vary with the type of reactor; we shall consider only emissions from PWRs and BWRs.

Most of the *gaseous* radioisotopes are beta-ray emitters and are therefore harmful only if inhaled or ingested. BWRs release the noble gas krypton-88, leading to a *maximum* annual dose to individuals of 0.001 rem per year, as well as small amounts of iodine-131 (0.0004 rem per year). The release of tritium gas, which quickly enters the water supply, is very low: a study of the Yankee Power Station west of Boston showed a highest dose of only 0.0002 rem per year. Carbon-14 is released by both PWRs and BWRs (0.0003 rem per year), mostly as radioactive carbon dioxide (CO_2) and methane (CH_4), which get into plant and animal food that is eventually eaten by people.

But all these releases are tiny compared to the natural background of 0.300 rem per year. Table 2 shows that, except for radioiodine, the doses from the airborne emissions of a PWR are typically lower than those from a coal-fired plant. The high bone dose from coal-fired plants results chiefly from radium-226, which is present in coal, as are radioisotopes of uranium, thorium, and lead. Fossil-fuel plants also release dangerous nonradioactive chemicals, such as sulfur and nitrogen oxides, which cause acid rain and depletion of the ozone layer. They also release particulate matter, which can cause lung cancer. None of these pollutants are released by nuclear plants.

Nuclear reactors also produce some *liquid discharges*. The most important is cesium-137, which has a half-life of 30 years (Appendix C) and is one of the major contaminants associated with the Chernobyl disaster. Cesium behaves chemically like potassium, being in the first vertical column in the periodic table (Appendix A), and is taken up by biological tissues. But, again,

Table 2. Maximum Radiation Doses Received from
Airborne Releases from 1000-Megawatt Power Plants
(mrem/year)[a]

Organ	Coal-fired plant	Pressurized-water reactor
Whole body	1.9	1.8
Bone	18.2	2.7
Lungs	1.9	1.2
Thyroid	1.9	3.8[b]
Liver	2.4	1.3

[a] From McBride *et al.* (1978—see Note 1). An *mrem* (millirem) is 1/1000 of a rem. Doses estimated at plant boundary.

[b] Assumes dairy cow on pasture at site boundary for entire year. The cow eats grass contaminated with radioiodine, which goes into its milk and is then drunk by humans.

these discharges are small. It has been estimated that anyone who swam in discharged coolant water continuously, night and day for a year, would receive only 0.00001 rem. Of course, some isotopes become concentrated as they move up the food chain, so fish near a reactor discharge would be more radioactive. Yet the Yankee Power Station study estimated that one would be exposed to only 0.0002 rem per year by eating such fish *every day*.

Finally, although the radiation dose limit per person set by international agreement is 0.5 rem per year, the U.S. Nuclear Regulatory Commission (**NRC**) allows a level only ¹/₅₀ of that—0.010 rem per year[2]—at the boundary fence of a nuclear power reactor. This level is never exceeded in the normal operation of modern plants. It corresponds to the additional dose of cosmic rays received by a passenger on a round-trip jet flight from San Francisco to Tokyo.

Transport

Nuclear materials must often be shipped long distances. Purified natural uranium is shipped to places like Oak Ridge, Tennessee, and Paducah, Kentucky, for *enrichment*, that is, increasing the relative concentration of uranium-235. The resulting *fuel*, which is not very radioactive, must then be shipped to nuclear power plants. After about three years in the power plant, the fuel is used up, or "spent," and is now *highly* radioactive, because of the fission products—it must be stored for some years on the site and then sent to either a reprocessing plant or a waste repository.

The radioactive spent fuel is shipped in casks designed to resist breakage (Figure 18). Bernard Cohen graphically described their properties:

> These casks have been crashed into solid walls at 80 miles
> per hour, and hit by railroad locomotives at similar speeds,

Figure 18. Radioactive spent-fuel elements are specially packaged before transport over public highways to storage sites. These pictures illustrate a "crash" staged to test the efficacy of the packaging. *Top*: A truck carrying radioactive material is struck by a train moving at high speed. *Bottom*: The truck is destroyed but the container for the radioactive spent-fuel elements, although dented, is basically undamaged. (Courtesy of Sandia National Laboratories, Albuquerque, New Mexico.)

without any release of their contents. These and similar tests have been followed by engulfment in gasoline fires for 30 minutes and submersion in water for 8 hours, still without damage to the contents. In actual practice, these casks have been used to carry spent fuel all over the country for more than 35 years. Railroad cars and trucks carrying them have been involved in all sorts of accidents, as might be expected. Drivers have been killed; casks have been hurled to the ground; but no radioactivity has ever been released, and no member of the public has been exposed to radiation as a consequence of such accidents.[3]

Clearly, the transport of nuclear materials is incredibly safe. In contrast, gasoline truck accidents kill about 100 Americans a year, and coal-carrying trains kill some 1000 people a year.[1] Yet New York City prohibits shipments of nuclear materials from Long Island through the city. New York City does not prohibit the travel of trucks or trains carrying gasoline or other dangerous chemicals, which pose a much greater hazard.

Explosion

Could a nuclear reactor ever explode like a nuclear bomb? Hardly. Reactor uranium is only 3% U-235, which would make a pretty poor bomb; nuclear bombs use uranium that is over 90% U-235. In addition, a reactor operates in just barely a critical condition, whereas a bomb explodes efficiently only by creating a highly supercritical mass. If a reactor does manage to go supercritical, as happened at Chernobyl, a nuclear explosion would start but would immediately stop, because expansion of the reactor core would quickly render the core subcritical. There would be a sudden increase in radiation and heat, which could cause melting or a relatively small explosion, as happened at Chernobyl. But a bomb-like nuclear explosion cannot occur.

Chemical explosions, however, can readily occur during a nuclear reactor accident. Hydrogen explodes in the presence of air and heat. Hydrogen may be produced by the reaction of coolant water with either the hot zirconium metal of fuel rods or the graphite moderator still used in some reactors. (All American commercial reactors use water for a moderator.) The small hydrogen explosion at Three Mile Island was contained by the containment building and had no environmental consequences. But it appears that a hydrogen explosion at Chernobyl, which had no containment structure, helped to blow the roof off the reactor building. The primary event at Chernobyl, however, was probably a steam explosion, produced by the reactor core going supercritical very quickly, producing a great deal of heat that converted the cooling water into steam.

Chernobyl was the only accident in civilian nuclear reactor history that resulted in a release of radioactivity with serious environmental consequences. But the damage was not caused by a nuclear explosion.

Sabotage

Although many people will concede that nuclear power plants are safe to operate, they feel that there is unusual risk in the possibility of sabotage. However, a little analysis will show that the situation is currently under excellent control.[4]

Sabotage, perhaps more accurately called terrorism, could involve two scenarios: either damaging the nuclear plant itself, or stealing nuclear fuel and making it into a bomb.

Nuclear power plants are typically highly guarded, being surrounded by multiple fences and open areas, backed up by electronic surveillance and constant policing. The danger of a direct assault seems extremely low. However, it is always possible that clever people might infiltrate a plant. If they were to damage the plant by actions in the control room, it is evident that they would have to know a good deal about the operation of a nuclear plant: they would find it hard to destroy the plant just by pressing buttons or throwing switches, any more than a naive person could manage the takeoff of a jet airliner.

Blowing up a nuclear plant with explosives, in order to contaminate the surroundings, is also not easy. The bombs used would have to be able to destroy the various containment shells around the reactor. Even if the saboteurs got into the reactor building, which would be very difficult, they would still need powerful explosives, which would be heavy and cumbersome.

In short, sabotage involving the nuclear plant itself seems highly unlikely. Nevertheless, it is clear that the industry must maintain vigilance, especially with regard to the entry of unauthorized persons, or anyone carrying a weapon.

Stealing bomb material from a federal weapons plant would also not be easy. People leaving a weapons plant must pass through a portal that will detect tiny amounts of radioactive material such as plutonium, even if the thief has swallowed it. Hijacking highly purified bomb-quality uranium or plutonium in transit would mean clashing with the armed guards of an armored truck and with the unmarked escort vehicle that follows it—both of which have radiotelephones.

However, *assuming* that several individuals were able to get their hands on bomb-quality material, what then? They would first have to transport this material to wherever they planned to make the bomb. Although the explosive material itself might not be excessively heavy, the shielding, primarily lead and concrete, necessary to prevent excessive radiation exposure would be heavy and cumbersome. The United States and Great Britain have jointly developed chemical processes that leave radioactive fission products in the plutonium, so that hijackers would be exposed to lethal radiation. Supposing that terrorists nevertheless managed to get such bomb material to a safe hiding place, they would then have to build a bomb. Now, we have all read about the clever high-school students who have figured out how to make a nuclear bomb, so you may think that it can't be all that hard. Planning something on paper, however, is very different from actually doing it. The U.S. Energy Research and Development Administration (ERDA), forerunner of the Department of Energy, has stated, ". . . a dedicated individual could conceivably design a workable device. Building it, of course, is another question and is no easy task."[5]

It would be far easier, far more reliable, and far less hazardous to the builders simply to explode a big TNT bomb. The bomb that blew up the American marine barracks in Beirut in 1983, killing 241 men, was quite powerful enough for the purposes of the terrorists. Indeed, there are far easier ways to kill *thousands* of people than to use a nuclear bomb, such as poisoning a city water supply or blasting open a large dam.

More likely than any of these scenarios for making a nuclear bomb would be stealing a completed weapon. However, all of the nuclear weapons in the U.S. Air Force are equipped with a permissive action link (PAL), which is a coded electronic switch that prevents anyone from arming a weapon without having received a secret arming code from a central command post. The U.S. Army and Navy have other safeguards but are likely soon to accept the PAL system, too.[6] (Of course, the stealing of completed weapons has nothing to do with the safety of nuclear power technology.)

In summary, it is not easy to take over a nuclear power plant, to steal nuclear bomb explosive and build an effective bomb, or to steal and use a completed bomb. One feels confidence in this conclusion from the observation that such acts of sabotage or terrorism have not so far even been attempted.

Breeder Reactors

A special type of reactor that has the remarkable ability to *produce more nuclear fuel than it uses up* is called a **breeder reactor**. How does it work?

The uranium fuel used in a typical power reactor is only 3% U-235, the remaining 97% being U-238, the common isotope of uranium. The U-235 is the active part of the fuel. It absorbs neutrons and splits apart, or fissions. The formula describing this process is shown in Figure 14. But the inactive U-238 can *also* absorb the neutrons produced by the U-235 fission. When this happens, the U-238 does not fission but is converted into other isotopes. For example:

$$^{238}_{92}U + 1\ ^{1}_{0}n \rightarrow\ ^{239}_{92}U \xrightarrow{\text{beta decay}}\ ^{239}_{94}Pu + 2\ ^{0}_{-1}e$$

In this reaction, U-238 is first converted to U-239 by the absorption of a neutron. U-239 then undergoes a *double* beta decay, in which two electrons are lost from the nucleus, thereby creating plutonium, with atomic number 94.[*] The plutonium-239 can undergo induced fission, so, like U-235, it is a good fuel for nuclear reactors. More important, it is an *excellent material for nuclear bombs*, partly because it releases more neutrons per fission than U-235.[7]

This production of plutonium-239 is called a **conversion reaction**, and it goes on all the time in ordinary reactors. In fact, after about 3 years, when the nuclear fuel of a reactor is exhausted, the fuel rods have had their U-235 depleted from 3% to 1%, but 1% of the U-238 has been converted to plutonium-239. Once a certain amount of plutonium has been made, one can build a breeder reactor, which typically uses plutonium surrounded by a "blanket" of ordinary uranium-238. We saw earlier that each plutonium fission yields up to three neutrons. At least one of these neutrons will be used to maintain the plutonium chain reaction. The other neutrons can be absorbed by the U-238 blanket to *produce more plutonium*, according to the reaction in the equation above. This kind of a reactor can be used, like any

---

[*]Note that the neutron has an "atomic weight" of 1, but a zero electric charge (atomic number). The electron has zero atomic weight but a negative charge, so its "atomic number" is -1.

reactor, to make electricity, but it will do this by burning plutonium at the same time that it produces, from U-238, even more plutonium than it burns. So the net effect is that *the breeder acts as if it were burning U-238*. Because U-238 is over 100 times as plentiful as U-235, breeder reactors would have enough fuel from uranium ore to last 10,000 years!

The production of plutonium from U-238 uses directly the fast neutrons produced by the U-235; there is no need to slow the neutrons down to thermal energies. Therefore, no moderator is needed. Such a reactor is called a *fast breeder*, meaning that it uses fast neutrons. Breeder reactors may use other conversion reactions, some of which in fact *do* require thermal, or slow, neutrons and therefore do require moderators. Such a reactor is called a *slow breeder*.

Breeder reactors cost more to build, but this cost is offset by the savings in uranium mining and processing, as all components of the uranium ore, both U-235 and U-238, are used. Breeder reactors could be the most effective reactor for the future.

8 Nuclear Reactor Accidents

Two aspects of nuclear power loom large in virtually any public discussion of this technology: accidents and waste disposal. These very emotional issues have generated a great deal of rhetoric and misinformation. Because of their prominence, I have devoted an entire chapter to each subject.

We established in the previous chapter that a normally operating nuclear power plant is very safe and does not contaminate the environment in a significant way. We also found that a nuclear plant will not explode like a nuclear bomb, and that the probability of sabotage is very low.

We will now discuss nuclear reactor accidents, looking first at the possibility of a meltdown. Then we will examine the three major nuclear reactor accidents to date: Three Mile Island in Pennsylvania, Windscale in England, and Chernobyl in the Soviet Union.[1]

Meltdown

The possibility of malfunction is the aspect of nuclear reactor operation that disturbs most people. In particular, they fear a reactor core **meltdown**, which, in its most dramatic description, would lead to the so-called *China syndrome*, where a reactor gets out of control and literally melts itself down into the Earth, presumably all the way to China. Could this happen?

First, we must examine the nature of meltdowns. If a *brand-new* reactor goes out of control and gets very hot very quickly, the control rods will usually drop into the core and immediately stop the chain reaction. If this is not done quickly enough, the core will expand and start to melt. The melting will result in a loss of criticality as well as a loss of moderator material, which together will stop the chain reaction. In either event, the melting will not proceed very far, and the consequences will be relatively minor. However, an *older* plant that has been operating for some time will have an accumulation of fission products in the fuel rods. In this case, even when the chain reaction has been stopped by the control rods, the heat from the radioactivity of these fission products may be sufficient to melt the core, unless emergency cooling is supplied. Furthermore, such a melted core will release dangerous fission products to the environment. Therefore, *it is the presence of fission products in the fuel rods that makes meltdowns more likely and more dangerous.*

With regard to the China syndrome, if the control rods failed, and the chain reaction took over, the reactor core would melt, and it might well melt itself down into the earth. But the core would probably go down no more than 20 or 30 feet; the core at Chernobyl did not go even that far. Once the core had melted and spread out, there would no longer be a chain reaction, and only the heat from the fission products would remain. There would remain a hot and horrible mess, which would be difficult to contain or clean up—again, as occurred at Chernobyl. But the China syndrome as such is a wildly exaggerated notion.

What are the chances of a reactor core meltdown? Meltdowns have occurred only twice in civilian nuclear reactors: at Three Mile Island and at Chernobyl. The Chernobyl accident had serious environmental consequences. At Three Mile Island, however, although 40% of the core melted, the pressure vessel that surrounds the core was not ruptured, and very little radioactivity was released. There are now some 450 civilian nuclear reactors in the world, and the first ones were built about 30 years ago. Assuming linear growth, we have therefore had an average of about 225 reactors operating for 30 years, which is to say that we have experienced roughly 6800 reactor-years, worldwide. During this extensive experience, there have been only two meltdowns. Does this mean that we can expect another meltdown in the next 3400 reactor-years, which would occur in only 7 years? Almost certainly not, for the reasons given below in our discussion of the Chernobyl accident.

In addition to the inherent safety factors that we discussed in the last chapter, nuclear power reactors have a large number of fail-safe devices built into their control systems. One of them ensures that, in an electric power failure, the control rods, which are held above the reactor core by electromagnets, will automatically drop into the reactor core, immediately stopping the nuclear reaction. Another is that an emergency core-cooling system (ECCS) is automatically turned on if the regular core-cooling system fails. In addition, as Daniel Ford, former executive director of the Union of Concerned Scientists, pointed out:

> A cardinal rule for the designers of commercial nuclear-power plants is that all systems essential to safety must be installed in duplicate, at least, so that if some of the apparatus fails, there will always be enough extra equipment to keep the plant under control. Federal regulations governing the industry require strict conformity to this prudent design philosophy.[2]

Three Mile Island[2]

In the early morning of 28 March 1979, a maintenance crew accidentally shut off the flow in the secondary water system at Unit 2 of the Three Mile Island nuclear power plant, near Harrisburg, Pennsylvania (*human error #1*). Normally, this action would not have created a problem. But in this case, it escalated into a near-disaster.

Emergency secondary-system pumps started up 14 seconds later (*proper safety function #1*), but during a routine test two days earlier, the valves in this secondary system had been shut but not opened up again (*human error #2*). The reactor operator on duty did not notice the red panel lights indicating this blockage (*human error #3*).

With no water flowing through the secondary system, the primary system—which cools the reactor core—heated up. This caused a pressure relief valve above the reactor core to blow open to relieve the pressure (*proper safety function #2*), accompanied by automatic dropping of the control rods into the reactor, which stopped the nuclear reaction (*proper safety function #3*). However, at this point, the residual heat from the fission products in the fuel rods was building up, which made it necessary to continue circulating water through the primary core-cooling system. The emergency core-cooling system (ECCS) pumps turned on within two minutes of the accident, thus cooling the primary system (*proper safety function #4*). Up to this point, there was still no real problem.

But, unknown to the operators, the pressure relief valve that had opened above the reactor core was stuck open (*equipment failure #1*), releasing steam into the containment building and leading the operators to think that there was too much water flow in the primary system. Therefore, they turned off the ECCS pumps shortly after they had started (*human error #4*). They also stopped the *regular* primary-cooling-system pumps (*human error #5*) and opened a drain line to remove even more water from the reactor (*human error #6*). Now the core became very hot, and much of the remaining water flashed into steam, preventing effective cooling of the fuel rods. Some of the fuel rods began to melt, releasing radioactivity into the cooling water. The pressure rose rapidly, keeping the pressure-relief valve open, and shooting more water and steam, now radioactive, into the containment building.

Many hours later, the operators managed to restore the cooling-water supply and get things under control, but by then, large amounts of water had been released into the containment building and had been pumped into storage tanks. The volume of water was so great that these storage tanks eventually overflowed, releasing a small amount of radioactive gas into the atmosphere. The top of the reactor core had been uncovered by water for several hours, and 40% of the core melted and broke apart, debris falling to the bottom of the reactor vessel. In total, 70% of the core was damaged.[3] Water in the core reacted with the hot zirconium-alloy fuel rods, producing hydrogen; this burned, but there was insufficient oxygen to cause an explosion. The pressure vessel and the containment building remained intact.

It is important to recognize that, although one major equipment failure did occur, this accident involved many *human errors*. The control systems of such a reactor are complex and difficult to operate. As Alvin Weinberg, former director of the Oak Ridge National Laboratory, noted:

> The accident at Three Mile Island was a prime example of both information deficiency and information overload. The deficiency lay in the failure of the operators to know that accidents almost identical, though less serious, had already happened at Davis-Besse and Rancho Seco. Had the operators known of these, they surely would have diagnosed their problem before the core melted down. Information channels were overloaded; once the accident started, the control room was deluged with a bewildering avalanche of lights, bells, announcements, data.[4]

Since Three Mile Island, great efforts have been made to train operators more thoroughly and to acquaint them with previously experienced problems. Great progress has also been made in providing operators with relevant data and with a clear analysis of these data, so that problems can be quickly diagnosed. Consequently, it is far less likely that such an accident will occur again. Indeed, to date, we have had no other major accident in U.S. civilian reactors. It must also be remembered that, in spite of the human errors at Three Mile Island, all but one of the fail-safe systems *did* work, preventing a complete meltdown.

No civilian was hurt as a result of the Three Mile Island accident. Yet this is hardly the impression that one got from newspaper and TV reports. There was a large release of the radioactive noble gas xenon-133 (half-life 5 days), amounting to 10 million **curies** (a unit of radioactive decay rate, or radioactivity—see Appendix B). The xenon exposure was of little biological significance, as noble gases are not taken up by human tissues—although the beta rays could have produced some slight exposure of the skin. The accident also released 15 curies of iodine-131. Such a quantity of this isotope would be dangerous if it were confined to a room, but injected into the atmosphere, it was quickly diluted to very low levels. The average dose from this release to individuals living within a 10-mile radius of the plant was only 0.008 rem[5]—equivalent to the dose of extra cosmic radiation that a passenger receives on a round-trip jet flight from Dallas to London. Some of the reactor personnel who had to shut things down, and later to clean things up, got fairly high exposures, but not beyond those considered acceptable for radiation workers. Two million people live within a 50-mile radius of Three Mile Island, and 325,000 of them are expected to develop cancer from natural causes in the next 30 years. It is estimated that the radiation released at Three Mile Island will add *one person* to this number.[6]

In terms of biological impact, Three Mile Island was certainly a minor accident. But the reactor core was left in a shambles, and many structures in the reactor building were contaminated with radioactivity. It took a decade to clean it all up, and the expense of this cleanup was horrendous. The result, then, was that Three Mile Island was primarily an *economic* disaster.

Perhaps even more important, it was also a *psychological*

disaster. The press went wild. The *Philadelphia Evening Bulletin* ran a three-part series, with headlines proclaiming "It's Spilling All Over the U.S.," "Nuclear Grave Is Haunting Kentucky," and "There's No Hiding Place." Helen Caldicott, a founder and former president of Physicians for Social Responsibility, wrote:

> [shortly] after the dreadful accident at Three Mile Island. . . We set up an office. . . staffed by Carol Belding, a mother who had been sitting with her baby on her knee during the accident at Three Mile Island.[7]

This vivid mother–child picture, with its chilling implications, shows the exaggerated reaction of even some professional people.

Windscale

The worst nuclear accident in the Western world happened in 1957, early in the history of nuclear reactors, at Windscale (now Sellafield) in Cumbria, England. It involved a gas-cooled graphite-moderated reactor designed to produce plutonium for the military. A faulty maneuver by an operator caused a fuel cartridge to split, releasing its contents, which then oxidized in the air, igniting the graphite moderator. The graphite burned furiously for almost two days. The fire was put out by flooding the reactor with water, and the whole antiquated system has now been sealed off.

The accident released 20,000 curies of iodine-131 into the atmosphere. This should have caused some 260 thyroid cancers in the exposed population—an area about 200 square miles around the plant—of which about 13 would be fatal. Because this is only a 1% increase over the natural level of thyroid cancer, it has not been statistically detectable.[8] The accident also released 240 curies of the uranium daughter isotope polonium-210 (see Figure 7). A recent report estimates that this exposure may have caused up to 33 people to develop other cancers, but as with the thyroid cancers, the percentage increase over normal levels of cancer are far too small to be actually detected.[9]

The Windscale accident was environmentally significant, but harm to the population was small. Three Mile Island was not environmentally significant and there was no harm to the population. I have dwelt on the details of Three Mile Island because of the great publicity, and the false perceptions, that surround this accident. The response to Windscale has been more rational, although military secrecy prevented the public from learning details until recently. Neither of these accidents was of great consequence. Only one nuclear accident in the world has been a true disaster: Chernobyl.

Chernobyl[10]

Even little children have learned a new word. Chernobyl!
  It is a new word for fear, and now our children wait to learn if an accident on the other side of the world will harm them.

DENIS HAYES, Chairman
Fund for Renewable Energy and the Environment, July 1986

On 25 April 1986, a series of events began at the Chernobyl power station, 60 miles north of Kiev, that would lead to the worst nuclear reactor accident in history. Thirty-one people were killed, hundreds were exposed to high levels of radiation, and thousands were exposed to low levels. 115,000 people were evacuated from the area and are only now beginning to return. Much of the surrounding land is still unfit for agriculture.

This was a very serious accident. However, because of the highly emotional reactions that surround such events, Chernobyl needs to be put into its proper perspective. With reference to the quotation above, for example: children in America are not even thinking about Chernobyl, and justifiably so, because they were not harmed at all.

The Accident

What happened was a series of incredible human errors resulting from an experiment that, though useful, did not have to be performed, and was in fact performed in a highly dangerous manner. The experiment was designed to determine if, in the event of a station blackout, the spinning generators would, while "coasting down," produce enough electricity to run the cooling pumps for a short time, until standby diesel generators could take over.

The events are diagrammed in Figure 19. The operators began to reduce power at the plant at 1:00 A.M. on 25 April. By two o'clock that afternoon, the power had dropped by 50%, and they shut off the emergency core-cooling system (ECCS) pumps (*safety violation #1*) because they did not want them to be activated during the tests. Then they waited until 11:00 P.M. to resume the experiment, because of the evening electricity demand, at which time they disengaged the automatic control system (*safety violation #2*) so they could control the system manually. Because the power dropped much lower than desired, almost all of the control rods were withdrawn in an effort to raise the power. At the low power at which the reactor was now operating, it was unstable and difficult to control manually. However, the operators then engaged two more coolant pumps (six were already operating) to make the test even more rigorous. This was *safety violation #3*, because so many running pumps could produce dangerous vibrations in the pipes. The sudden spurt of water from the extra pumps cooled the system and reduced steam pressure to a level where the reactor would automatically signal an emergency and would shut itself off. To prevent this, the operators switched off this warning system (*safety violation #4*). The reactivity of the reactor was so low at this point, and the reactor was so unstable, that a computer warned the operators to shut down the reactor immediately. By now committed to finishing the test, they did not shut down the reactor (*safety violation #5*).

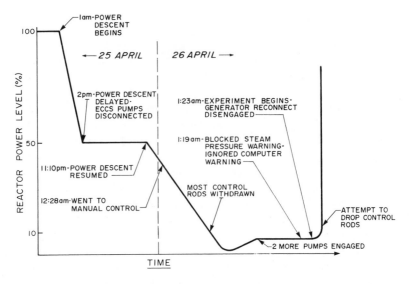

Figure 19. Graph of the major events leading to the Chernobyl accident. The time axis is not to scale.

At 1:23 A.M. on 26 April, the "coasting down" experiment finally began. There was still another safety feature operative, designed to automatically reconnect the generators to the reactor if the reactor power suddenly surged. The operators shut this automatic reconnect off (*safety violation #6*) because they wanted to be able to repeat the experiment if it didn't work the first time.

The pumps were then connected to the generators, and as the generators slowed down, water flow through the reactor core dropped. This was a boiling-water reactor, in which some of the water boils as it passes through the core. With a lower water flow, more steam was generated, causing a rapid surge in the power of the reactor. This extra power meant extra heat, which in turn produced more steam, in a sort of chain reaction. Within 3 seconds, the power went so high that the shift manager tried to shut it down by dropping the control rods into the reactor core. But it was too late: the reactor was already out of control. In the next 4 seconds, the power surged to 100 times the reactor's capacity. The uranium fuel disintegrated, burst through its cladding (the tubes that contain the fuel), and contacted the cooling water. Enormous steam pressure developed that blew the steel cover off the reactor and blew out the roof of the building. This hot steam probably also reacted chemically with both the zirconium alloy of the fuel rods and the graphite of the moderator, releasing hydrogen into the air above the reactor, which caused a hydrogen explosion within the next few seconds. These steam and hydrogen explosions destroyed most of the building, and immediately killed two people. Highly radioactive fission products and burning blocks of graphite were strewn about the plant.

Radioactive debris from the reactor was blown skyward. Later, fission products in the core kept the reactor hot and the graphite burning, so more radioactive material was thrust upwards over the next 10 days. Up to 10% of the radioactive material in the reactor was ejected into the atmosphere. Much of this fell to the ground within 20 miles, but some was wafted by high-altitude winds to places as distant as England (Figure 20). It is remarkable that the city of Kiev, only 60 miles to the south, received relatively little radiation because it was upwind of Chernobyl throughout most of the accident.

In the weeks after the accident, tons of boron carbide, dolomite, lead, and clay were dropped from helicopters into the reactor core, which is now a quiescent (but still warm) lump of material that continues to be a radiation hazard for anyone spending much time close to it. One assumes that eventually this material will be removed and buried in a deep repository.

There are many reasons why the Chernobyl reactor accident occurred and escalated into such a dangerous situation. Most of

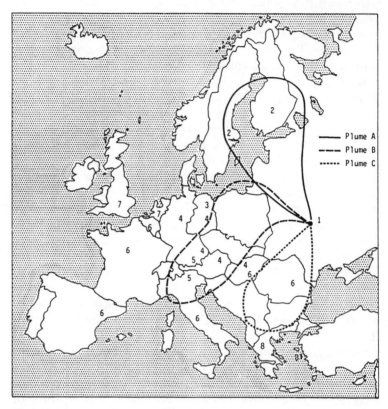

Figure 20. Areas immediately affected by radioactive fallout following the 1986 Chernobyl nuclear reactor accident. The numbers on the map indicate the day after the accident when fallout arrived. The pattern swept from high in the north on the second day, across Sweden and Finland, in a counterclockwise motion down to Greece on the sixth day. (Courtesy of United Nations Scientific Committee on the Effects of Atomic Radiation [UNSCEAR], *1988 Report to the General Assembly*—see Bibliography.)

these things could not happen in, or apply to, the civilian reactors of Western countries:

1. This was not an excusable accident, like that at Three Mile Island, where equipment failed and the operators did their best to solve the problem. Chernobyl was a *deliberate and dangerous experiment* that went awry.

2. The operators committed *six safety violations*. Even the Soviets characterized this as "unbelievable." Had any one of these not occurred, either there would have been no accident, or it would have been minor. The decisions made at Chernobyl reflect very poor judgment on the part of the operators, who, during this experiment, were being guided, not by nuclear engineers, but by electrical engineers more familiar with turbine generators than with reactors.

3. In almost all reactors, including that at Chernobyl, the power decreases with rising temperature, but the Chernobyl reactor had the unusual property that power *increases* with rising temperature below about 20% of full power—a highly unstable situation that can readily lead to a runaway chain reaction.[11]

4. The Chernobyl type of reactor operates at lower water pressure than American graphite reactors (used only for military purposes). This lower pressure permits the water to start boiling when it is only one-third of the way down the cooling tubes; thus it is much easier to make *all* the water boil. When this happens, steam pressure can blow apart the water tubes, leading to the production of highly explosive hydrogen.

5. Had the plant not had a graphite moderator, there would have been little fire in the reactor core. Graphite can burn with a hot glow, like coal, for a long time. The graphite fires at Chernobyl and at Windscale accounted for much of the radioactivity going into the atmosphere, and these fires severely retarded efforts to deal with the accidents. The graphite moderator at Chernobyl also contributed to hydrogen production, and hence

to the second explosion, which blew even more debris into the air.

6. The reactor was not in a full-fledged containment building, which could have served as a barrier to the release of radioactive material.

As noted above, most of these things would not apply to Western reactors. None of them would apply to American civilian pressurized-water reactors, and only Item 6 might apply to some of our boiling-water reactors, whose containment is not complete.

The Fallout[12]

Three groups of people were affected by the Chernobyl accident: (1) those in or near the reactor building, (2) those within about a 20-mile radius of the Chernobyl plant,[13] and (3) those outside the Chernobyl area, in Europe and the western USSR.

At the power plant itself, 237 firefighters and plant workers were affected by fire and/or radiation, with 128 exposed to radiation doses from 80 to 1600 rems. A total of 31 people died, two within the first 12 hours from mechanical injury and burns. The other 29 were mostly firefighters, most of whom died within 3 weeks from high doses of radiation (over 400 rems). Dr. Peter Gale, an American physician, gave bone-marrow transplants to 13 of these victims but saved only 2. A year later, only 13 of the roughly 200 survivors were still invalids and unable to return to work.

The second group affected were the people living in the Chernobyl area. Approximately 50,000 got doses averaging 50 rems. This is the maximum permissible exposure for a 50-year-old radiation worker. Several hundred of these people will probably die of premature cancer over the next 50 years.[14] All of the 115,000 people living within about 20 miles of the reactor were evacuated. By the mid-1990s, it may be possible to measure, within this evacuated population, increased incidences of leukemia, which is the most likely cancer to result from irradiation. For other cancers, the increased death rate will be so low that it will not be possible to determine whether it was caused by Chernobyl fallout. However, as of February 1989, the 1950 children born to women in the area who were pregnant at the time of the explosion *were all normal*, although they are still being watched for signs of leukemia.[15]

The third group includes people in Europe and the western USSR, outside the Chernobyl area, who have received and will receive low doses over a long period of time. Most of the radioactive release was in short-lived isotopes, such as iodine-131. Because of its short half-life of 8 days, I-131 radiation ceased within a few months after the accident. The radioiodine in the air and in food should already have induced some thyroid cancers, but these are highly curable. More serious is the 10% of the radioactive release that was in the form of long-lived isotopes, such as cesium-134 and -137 (see Appendix C). The major problem is with cesium-137, which has a half-life of 30 years. One million curies of this isotope were released. Half of the danger will result from direct gamma radiation from cesium on the ground in the early years, and half will come from cesium in food during the later years. It is difficult to estimate the doses of radiation from any fallout outside the 20-mile zone around Chernobyl. Although atmospheric measurements were made in Europe, they are spotty[16], and give only a rough idea of the radiation levels. People in the northwest Ukraine and Byelorussia will probably receive less than 10 rems, while people outside this area should average less than 0.075 rem (Figure 21), which is only one-quarter of the annual background dose.

American researchers have estimated that Chernobyl fallout will cause 17,000 deaths from premature cancer, meaning that, for 17,000 people, death up to 10 years sooner than normal will occur over the next 50 years.[14] Of these deaths, 37% will occur in the USSR, 60% in non-Russian Europe, and 3% in Asia outside the USSR, with virtually no impact on the population of the United States.[17]

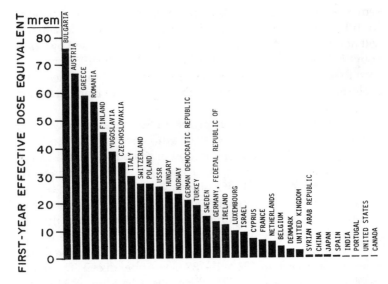

Figure 21. Countrywide average first-year committed effective dose equivalents (see Appendix B) from the Chernobyl accident. The highest dose (in Bulgaria) was only 25% of the background level of 0.3 rem (300 millirem). The USSR value was low (between those of Poland and Hungary) because it was averaged over the vast area of the Soviet Union. (Courtesy of United Nations Scientific Committee on the Effects of Atomic Radiation [UNSCEAR], *1988 Report to the General Assembly*—see Bibliography.)

It must be emphasized that these are *estimates*, not measurements; they could easily be off by a factor of 2 or more. For example, Swedish researchers have predicted a total of only 6000 increased cancer deaths due to Chernobyl,[18] and the European Community estimates only 1000 extra cancer deaths in its population over the next 50 years—among 30 million normal cancer deaths.[19] Another important point is that we are speaking of very large populations. Almost any estimate of risk in a large population will look bad when expressed as numbers of people who will die. For example, it is estimated that 1000 people in America will die prematurely from cancer every year just because of the radioisotopes normally present in their bodies. A thousand people sounds pretty awful, but it is a tiny fraction of the U.S. population: 7000 Americans die every year from falls.

These 17,000 cancer deaths over 50 years are immersed in a sea of 123 *million* normal cancer deaths that will occur during this period in Europe and the western USSR,[14,17] so that this is *an increase over normal cancer deaths of about $1/100$ of one percent, which is impossible to measure*. Such a slight increase is no greater than the risk one would expect from increased cosmic rays in a population living at a high altitude, as in Denver, or among businesspeople who travel frequently on airliners. (Neither of these increases, of course, has been actually observed because the percentage increases are too small.) Among the 600 million people living in Europe and the western USSR, radon in homes alone is estimated to produce about 10,000 cancer deaths *every year*. In this same population, there will be 1 million cancer deaths *per year* just from smoking. The estimated Chernobyl cancer toll of 17,000 over 50 years amounts to only 340 extra cancer deaths per year, or $1/3000$ *the death rate due to smoking*.

It is also important to recognize that, even though *dose rates* (the dose per unit time) rose in some parts of Poland and Scandinavia to as much as 40 times background levels, they declined rapidly after a week. Consequently, the increased *total dose per year* was relatively small outside the Chernobyl area. Indeed, the average *lifetime* radiation exposures of Chernobyl evacuees—who got the highest doses—is no greater than the *yearly* exposure of all of us to background radiation (Figure 22).

Another consideration is that the radiation from Chernobyl will produce "late cancers," the average induction period for cancer being about 25 years.[20] Most of the victims of Chernobyl fallout will die 1 to 10 years sooner than normal. Like smokers who die from lung cancer and people who die from falls, they will be older people, who will lead fairly long lives before succumbing.

Table 3 compares Three Mile Island and Chernobyl with

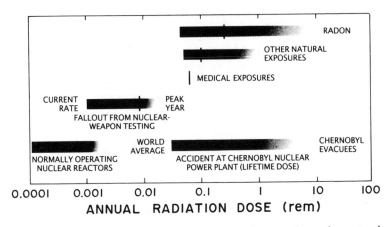

ANNUAL RADIATION DOSE (rem)

Figure 22. Range of annual radiation exposures due to radon, other natural exposures, and medical radiation, compared to annual exposures due to fallout from atmospheric nuclear weapons testing (which virtually ceased in 1963) and normally operating nuclear reactors, and the *lifetime* exposure of evacuees from the Chernobyl power plant accident. Short vertical lines indicate average indoor levels in the U.S., except that the line for nuclear weapons testing is a world average. Note that the horizontal scale is logarithmic. (From Anthony V. Nero, Jr., "Controlling indoor air pollution." Copyright 1988 by SCIENTIFIC AMERICAN, Inc. All rights reserved.)

other sources of radiation exposure. The doses to people living close to the reactor were indeed high at Chernobyl. But the dose to the world population was equivalent to only three years of airplane travel, or one month of background radiation. It was only 2% of that produced by the U.S. and Soviet testing of nuclear weapons in the atmosphere during the period 1950 to 1963.

Chernobyl was vastly different from other reactor accidents. In terms of fallout from the radioisotopes iodine-131 and cesium-137, Chernobyl was 2000 times worse than Windscale, and 500,000 times worse than Three Mile Island. At Three Mile Island, only 15 curies of radioiodine were released. At Chernobyl, about 100 million curies of dozens of isotopes were re-

Table 3. Estimated Doses to Populations[a]

	Area affected	Population	Average dose (rem)
Three Mile Island	50-mile radius	2 million	0.002
Chernobyl	20-mile radius	130,000	12.
	Worldwide	3 billion	0.02
Atmospheric testing	Worldwide	3 billion	1.0
Airplane travel	Worldwide	3 billion	0.006[b]
Background radiation	Worldwide	3 billion	0.200[b]

[a]From Table 2, Henry D. Royal, 1990, "The Three Mile Island and Chernobyl reactor accidents," Chap. 17, in Fred A. Mettler, Jr., Charles A. Kelsey, and Robert C. Ricks (Eds.), *Medical Management of Radiation Accidents.* Boca Raton, FL: CRC Press.
[b]Per year.

leased. At Three Mile Island, no one was hurt, whereas 31 people died at Chernobyl, 200 more were badly irradiated, and perhaps 17,000 will die prematurely of cancer. Three Mile Island was an economic and psychological disaster but cannot even be considered an accident from a biological standpoint. Chernobyl, on the other hand, was a serious accident, economically *and* biologically. We do not want to have any more Chernobyls. But it is most unlikely that we shall, in view of the phasing out of Chernobyl-type reactors, the superior design of Western reactors, and the development of new "inherently safe" reactors.

It is of critical importance that the public have an accurate perspective on nuclear power. Many people have argued for the demise of nuclear power because of Chernobyl, which killed 31 people. We should remember that *Chernobyl is the only civilian nuclear reactor accident that has killed anyone, anywhere in the world*. The Japan Air Lines crash of a Boeing 747 nine months earlier that killed 520 people was not followed by a cry to stop commercial air travel. The accidental release of the deadly chemical methyl isocyanate in 1984 at a Union Carbide plant in Bhopal, India, killed at least 3700 people outright and injured 30,000 more. Yet there is no move to stop the building of chemical plants. Our perceptions of danger seem to be badly skewed.

Inherently Safe Reactors

On the morning of 3 April 1986, three weeks before Chernobyl, a small nuclear reactor in Idaho, operating at full power, suddenly underwent one of the most alarming events in a nuclear reactor operation: the cooling fluid stopped flowing. However, the engineers merely sat quietly and watched, waiting to see what would happen. The temperature of the coolant rose to 1300° F, but it didn't boil, and it continued to cool the reactor. The nuclear reaction stopped, and the coolant temperature dropped within five minutes to its normal 900° F.

What was going on? This was a test conducted by Argonne National Laboratory of a new type of experimental breeder reactor, EBR-II, which is an **inherently safe reactor**. In such a reactor, a meltdown upon loss of the core coolant is almost impossible, even if the operators do absolutely nothing. Clearly, such reactors hold great promise for the future of nuclear power.

However, before we discuss the exciting prospect of a new breed of inherently safe reactors, let us look more closely at the safety of our present reactors.

Estimates by the Oak Ridge National Laboratory and related groups show that the typical U.S. reactor before Three Mile Island had about 1 chance in 1000 of a meltdown, either complete or partial, per year.[21] After the incorporation of improvements mandated by the Three Mile Island incident, however, this estimate rose to about 1 chance in 10,000 of a meltdown, meaning that we can expect a meltdown among the 100 U.S. reactors only once every 100 years. These figures also indicate that, among the world population of 450 reactors, we may expect one meltdown every 22 years. This might be acceptable, for *probably only one in four meltdowns would involve a significant exposure of human populations to radioactivity*, so that such population exposure would happen only about once every century.

In short, *the present generation of reactors is mostly very safe.* Few graphite reactors are still in use. None are used in the U.S. for commercial power. About one-third of American reactors are boiling-water types (BWRs) that have less effective containments than the pressurized-water types (PWRs). However, this fact was considered in the above estimate that one in four meltdowns would seriously expose the general population. The Soviet Union is currently modifying all of its Chernobyl-type reactors and plans to phase them out.

Nevertheless, the possibility of a Chernobyl-like accident, even once a century, is still cause for concern. Certainly, the public fears a similar accident. One answer to these fears would be to use inherently safe reactors. How do they work?

The EBR-II reactor mentioned above is cooled by liquid sodium instead of water. Liquid sodium has a boiling point of 1650° F and will continue to cool even a very hot reactor without boiling. In American water-cooled pressurized reactors, a loss of pressure immediately causes the superheated water to boil, to convert to steam, and, as a result, to lose most of its cooling ability. Also, sodium is an excellent heat conductor, so if a reactor core gets unusually hot, the heat is spread uniformly and the whole core expands. This expansion moves the uranium atoms farther apart, so that the rate of nuclear fission drops and the core cools down.

Another new design is the PIUS (Process Inherent Ultimately Safe) reactor currently being developed in Sweden. In this reactor, the entire core, as well as the primary-cooling-system pumps and the heat exchangers that produce steam, are surrounded by water that contains the metal boron, which is an excellent absorber of neutrons, for which reason it is commonly used in control rods. If a pipe or valve failure should interrupt the normal flow of coolant, the lowered water pressure would immediately allow this borated water to flow into and around the reactor core, thus shutting down the nuclear reaction. This large volume of water would keep the core cool for several days, without the intervention of operators or of emergency cooling systems. Such a system would also be proof against earthquakes, and against sabotage by explosives. It is of some interest that the Shoreham nuclear reactor on Long Island, aborted because of public outcry, has a boron tank that would have flooded the reactor core in the event of a coolant failure.

A third design, which has been operated and tested in Germany, is a high-temperature gas-cooled (HTG) reactor, whose inherent safety comes from its small size and low power density. It produces only 3 kilowatts of power per liter of core volume, whereas an American pressurized-water reactor produces 100 kilowatts per liter. If coolant is lost, the nuclear chain reaction is terminated after a modest rise in temperature of the reactor core. The HTG reactor is cooled with helium, a chemically inert gas, which could not burn or explode as the hydrogen, generated by water, did at Chernobyl and Three Mile Island. Although these reactors are smaller than the American type, several of them together would generate equivalent power.

Such inherently safe reactors, as one might guess, are more expensive to build than the usual kind. However, because of their inherent safety, they might not require containment buildings, and this saving alone would probably offset the additional cost of the reactor. In addition, because of the inherent safety, the Nuclear Regulatory Commission construction requirements could be greatly relaxed, so that a reactor could be built perhaps in 6 years instead of the present 12. The time it takes to build a reactor in the United States—and the consequent great cost—is one of the major factors that inhibits nuclear power development in this country.

In summary, most of the reactors in the world are now quite safe, and they are constantly being improved. In addition, we now have the ability to build reactors in which a meltdown is virtually impossible. Clearly, the operation of nuclear power plants can be made as safe as the world desires.

9 Nuclear Waste Disposal

Problems associated with nuclear power are often magnified in the public eye. But the problems of nuclear waste disposal are in fact complex and serious. These problems are being solved, but their solution requires ingenuity and commitment.

It will help at the outset to develop a few perspectives about nuclear waste. First, we must remember that the earth, the oceans, the atmosphere, and even ourselves are radioactive. The oceans and lakes of the world have a total radioactivity of 500 billion curies, due mostly to potassium-40, the most common radioisotope in our own bodies. Some of the radioactivity of the oceans comes from its 4 billion tons of uranium, along with uranium decay products, which include radium. The atmosphere has 60 million curies of tritium,[1] the radioactive isotope of hydrogen, which is produced by cosmic rays. Most of this tritium is present in water vapor, some of which enters the oceans. The rocks of the continents are virtually all radioactive, containing uranium, thorium, and radium; in fact, it is from these rocks that we get our uranium ore for nuclear reactors and bombs.

Doesn't the nuclear power industry add to this background radiation we already have? Yes and no. Over a time scale of billions of years, the nuclear industry actually *reduces* the world's radioactivity, by "burning up" uranium. But in the short run it adds to the radioactivity, because it converts long-lived isotopes like uranium-235 (half-life = 700 million years) to shorter-lived isotopes, like plutonium-239 (24,000 years—see Appendix C) and cesium-137 (30 years). These shorter-lived isotopes have higher radioactivity, emitting more radiation per hour.

Natural background radioactivity is a hazard, but only a slight one. Because of the tremendous dilution of radioisotopes by the great volume of nonradioactive rock, ocean water, and atmosphere, background radiation is relatively small and not particularly harmful. Biological organisms have lived with it for billions of years. *Dilution is the important factor.* For example, the radioactive waste (hereafter called **radwaste**) produced by uranium mining and milling releases radon into the air. Yet, it has been estimated that, even if all the world's electricity were made by nuclear reactors, and if all the old mine and mill tailings were left exposed to the air, 1 year of uranium mining would add only 1 part in 6000 to the natural radon content of the atmosphere.[1] The small releases of radioactive gases from normally operating nuclear power plants are quite harmless, because the gases are quickly and effectively diluted. And if defective containers of radwaste slowly leak their contents into the atmosphere, the oceans, or even groundwater, dilution will often take care of the problem.

One argument against such *dispersal* of radwaste is that biological organisms may concentrate a radioactive substance, and the degree of this biological concentration may increase greatly as one goes up the natural food chain. Organisms frequently "hoard" certain atoms and molecules, as the human body hoards every bit of iron and iodine it can get. Thus, the insecticide DDT is concentrated 1 million times in passing from water all the way up the food chain to fish-eating birds. Radioactive phosphorus, zinc, iron, and iodine may be concentrated in seafood to a level tens of thousands of times greater than that found in seawater. Dispersal is therefore not generally favored as a means of radwaste disposal.

We have mentioned that short-lived radioisotopes show

higher radioactivity. This may be described in the following very important relationship:

$$\text{Activity is proportional to: } \frac{1}{\text{Half-life}}$$

where **activity** means the degree of radioactivity, usually measured in curies. This equation states that a radioactive isotope that has a very long half-life, like uranium or plutonium, will have a very low activity level, or in other words that the *rate of decay* will be very slow. Thus, one can hold a piece of uranium ore in one's hand, as uranium miners do, and receive only a very low dose of radiation. Conversely, radon gas, with a half-life of only 4 days, has a very high activity and is dangerous even in small amounts. One can understand this equation by recognizing that a piece of radioactive material has a fixed number of atoms in it that are going to decay. It will either "burn very brightly for a short time" or "burn very slowly for a long time," just like anything else that burns up.

The important consequence of this equation for radwaste disposal is that many radioactive substances that have a very long half-life are not very radioactive at all. When the public is told about isotopes that have half-lives of millions of years, they often assume that these isotopes must be very dangerous. But this will be true only if these isotopes are *highly concentrated*. Although such high concentration does exist for nuclear fuel and fission products, it does not exist for many kinds of radwaste.

In addition to the physical half-life of a radioisotope, we recognize what is called the *biological* **half-life**, which is a measure of the time that an atom or molecule is retained in the body. Living things excrete or "flush out" most atoms and molecules relatively quickly, in urine, feces, sweat, or exhaled breath. But some other things, like iron or iodine, may be retained for a long time. Note that the biological half-life is a *chemical* property: it is the same for all isotopes of the same element—such as all isotopes of iodine—whereas the physical half-life is different for each radioisotope—such as I-131 and I-129. An isotope with a *very short biological half-life* (such as tritium—biological half-life of 1 week) will be flushed out of a living system quickly, before it can do much damage. An isotope with a *moderately long biological half-life* (for iodine, 5 months) will be dangerous only if it also has a short physical half-life (for ^{131}I, 8 days), because it will then completely decay before leaving the body. ^{129}I, on the other hand, is not dangerous, because its physical half-life is 17 million years, and it is gone from the body before it has hardly begun to decay. Isotopes with a *very long biological half-life*, such as strontium or plutonium, which settle in bone and stay there for decades, are usually dangerous regardless of their physical half-lives. We deal here mostly with the physical half-life, and will always mean this unless we specify otherwise.

Finally, we need to consider different levels of radwaste. **Wastes** are classified as either *low-level* or *high-level*. These terms are defined in different ways, but they should refer, not simply to the concentration of radioactivity, but to the *long-term health hazard after disposal*. Thus, "true" low-level wastes could contain high concentrations of isotopes with short half-lives (e.g., iodine-131) or low concentrations of isotopes with long half-lives (e.g., uranium mining wastes). "True" high-level wastes should include relatively high concentrations of things with half-lives longer than about a year, including **spent fuel** (used fuel, which contains isotopes like cesium-137 and plutonium-239), reprocessing wastes, and parts of dismantled power plants.

Unfortunately, some dangerous types of waste, such as cooling-water filters and old control rods from nuclear power plants, have been classified as "low-level" waste and have thus escaped rigid control. This unfortunate classification has created much concern among the public, to say nothing of danger to them. It is a political problem and can be solved simply by defining as high-level waste all wastes that pose a long-term health hazard. Because the problems associated with "true" low-level waste are relatively minor, I do not discuss them in this book.

Sources of Waste

Dangerous radwastes come primarily from *nuclear reactor technology*, which makes it possible for us to have civilian power reactors, military reactors for producing bombs, and reactors for nuclear-powered ships.

Other sources of radwaste exist, but generally create far less hazard. They include the radioisotopes used in nuclear medicine and in biological research, as well as the clothing, gloves, and other apparatus used in such work. Industry produces some radwaste, as from mineral exploration. Most of these other uses result only in the production of low-level waste and do not pose a serious hazard to the public.

Some Earth satellites are powered by small electric generators that use the heat produced by nuclear decay or nuclear fission. They could be dangerous if used in large numbers, but a single accident may pose little hazard. For example, one year after a Soviet reconnaissance satellite containing 100 pounds of U-235 burned up in the stratosphere in February 1983, it was just barely possible to measure the increase of stratospheric U-235 over the background level.[2] Nevertheless, the continued use of such power sources in Earth satellites does present an unnecessary hazard and should be terminated. This is particularly true of satellites that use plutonium-238 (not to be confused with the fissionable Pu-239). Pu-238 emits strong alpha and gamma rays, and has both biological and physical half-lives of about 100 years. As the Federation of American Scientists recently stated:

> . . . there is approximately a 15% failure rate for past space nuclear power missions both US and Soviet. Arguably, the worst accident to date was the 1964 reentry and burnup of the US SNAP9A, which tripled the world's environmental burden of plutonium-238.[3]

Although we should undoubtedly stop using nuclear-powered *satellites* in Earth orbit, where they pose a continuous threat, this concern need not extend to nuclear-powered *deep-space vehicles*, which will pass through the Earth's atmosphere only once. The public hue-and-cry over the launch in October 1989 of the Galileo space mission to Jupiter, which carried a Pu-238 power source, was sadly misplaced.[3]

It is not generally recognized that *only 10% of our high-level nuclear waste has been produced by military activity*. Most of it is produced by civilian nuclear power technology. Yet this relatively small military production has created most of our high-level waste-disposal trauma. The recent revelations of irresponsible handling of waste at our nuclear-weapons production plants are shocking, but must not be confused with the excellent record on waste management of the civilian nuclear-power industry.

Disposing of obsolete nuclear weapons is also part of the high-level waste-disposal picture. However, there are two nice solutions to the problem: (1) the nuclear materials in old weapons can often be recycled into new weapons, and (2) the plutonium and uranium from obsolete weapons can be used as fuel in nuclear power plants. The latter is a kind of modern version of the "beating of swords into plowshares."

Types of Nuclear Reactor Waste[4]

There are four categories of radwaste associated with nuclear reactors, whether military or civilian. Some properties of the isotopes falling into these categories are shown in Appendix C.

1. **Transuranic atoms** are elements like neptunium, plutonium, and americium, which are produced in reactor fuel upon absorption of neutrons by uranium nuclei. They generally have long half-lives, of thousands of years. They emit nonpenetrating alpha particles and

weak gamma rays, and are therefore *dangerous only when inhaled or ingested*. The transuranics constantly decay into uranics.

2. **Uranics** are uranium, thorium, and their decay products, such as radium, radon, and lead. They also are primarily alpha-particle emitters. Uranium is the primary fuel for nuclear power, so it is a major component of spent fuel waste. Like the transuranics, most uranium isotopes are relatively benign unless inhaled or ingested. Radium, a decay product of uranium, is a great problem, because it decays to radon-222 gas. This radon isotope has a short 4-day half-life, but it decays into high-activity daughter particles, like polonium-218 and -214, that may be deposited in lung tissue.

3. **Fission products** emit the poorly penetrating beta rays, and are thus hazardous only if inhaled or ingested. The most important ones are strontium-90 and cesium-137, both with physical half-lives of about 30 years. These two isotopes were the most significant components of the Chernobyl fallout. Sr-90 behaves like calcium: it gets into milk and eventually into human bone, where it may stay for years. Cs-137 behaves chemically like potassium and gets into all body tissues, but its biological half-life is only about 3 months. Xenon-133 (half-life 5 days) and krypton-85 (half-life 11 years) have been released in large quantities in reactor accidents. However, being noble gases, they are chemically unreactive and do not stay in biological tissue—they just irradiate the skin, or the surface of the lung if they are inhaled. Iodine-131 (half-life 8 days) is much more dangerous. Although it is mostly excreted at once in the urine, what is left accumulates in thyroid tissue with a biological half-life of about 5 months, so it stays there until it has all decayed.

4. **Neutron bombardment products** mostly have half-lives of only a few years and are primarily beta-emitters, although cobalt-60 and niobium-94 emit strong gamma

rays. These products do not originate in the fuel itself; they are produced by neutron bombardment of surrounding metals, which make up fuel rods, control rods, and other core structures. Tritium, the heavy isotope of hydrogen, is produced by neutron bombardment of cooling water. Except for the tritium, these products become a problem only when a nuclear power plant is being dismantled.

The Nuclear Fuel Cycle

Figure 23 diagrams the commercial nuclear fuel cycle, from the mining of uranium ore to geologic disposal. The solid arrows indicate existing steps and the broken arrows steps that have not yet been widely implemented. The stages of interest for waste disposal are (1) *milling* of the uranium; (2) interim storage, or *holding*; and (3) **geologic disposal** either of spent (used) fuel that has not been reprocessed (right-hand dashed arrow), or of high-level waste from a reprocessing plant. Another problem is created by (4) *dismantling* of old power plants and weapons plants. We discuss now the problems of milling, holding, reprocessing, and dismantling. Geologic disposal is then discussed in some detail.

Milling

Unlike coal that is mined and directly used, uranium ore goes to a mill that extracts most of the uranium, leaving behind a tremendous volume of **mill tailings**. Extraction of the fuel for a large 1000-megawatt reactor, operating for one year at 80% capacity, leaves 72,000 cubic meters of mill tailings[4]—a volume that would cover a football field to a height of 100 feet. This waste actually has a lower radioactivity than the original ore, so it is really low-level waste. Problems arise only because these wastes have often been handled irresponsibly, which has re-

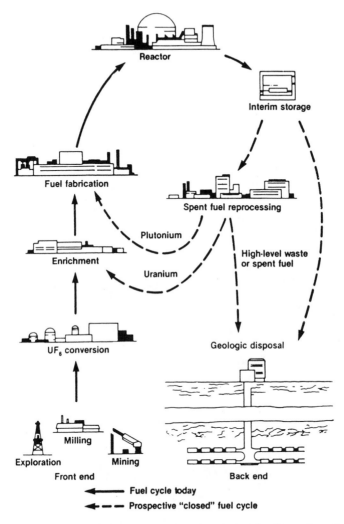

Figure 23. Diagram of the commercial nuclear fuel cycle. (From Council on Environmental Quality data in *Managing the Nation's Commercial High-level Radioactive Waste*, March 1985, OTA-0-171. Reprinted with permission from Office of Technology Assessment, U.S. Congress, Washington, DC.)

sulted in extensive release of radon-222. For example, mill tailings have been used in Grand Junction, Colorado, as landfills in the construction of homes and schools! Clearly, this was not a technical problem, but a problem of social responsibility. These tailings can simply be buried several meters deep, perhaps in old mines, where they will pose less hazard than the original uranium-containing earth.

Holding

The important wastes produced by nuclear reactor technology are *reactor wastes*. These consist primarily of spent fuel and high-level wastes from the reprocessing of spent fuel. In both cases, much of the radioactivity is in short-half-life isotopes, which produce a great deal of radiation and heat. Therefore, these wastes are usually stored for many years in a *cooling pond* of water, which conducts heat away from the waste material. This water also completely absorbs the neutrons, so that one may look into such a pond without being significantly exposed to radiation. Such storage for 40 years will reduce the radioactivity of spent fuel to about $1/100$ of its original level,[5] which makes it far easier to handle for long-term disposal.

One may think that 40 years of "holding" is a ridiculously long time to have the stuff hanging around! Not so. The *easiest* way to deal with fresh waste is simply to put it into a cooling pond right at the site where it was produced. The volumes of waste are small, and the ponds do not have to be large. Not only does this storage minimize handling, but it keeps the waste away from the public during the period when it is most dangerous.

Reprocessing

The fuel for a nuclear reactor must be replaced after about three years, by which time the U-235 concentration has been depleted from 3% to 1%. In the meantime, however, neutrons from the fission process have been slowly "converting" about 1% of the uranium-238, the major component of the original fuel, into plutonium-239, as described in our discussion of breeder reactors in Chapter 7. Thus, a spent fuel rod contains, among other things, 1% U-235 and 1% Pu-239. These are both fissionable isotopes that could be used in reactors, if they were removed and concentrated. That is precisely what is done in a **reprocessing** plant. About 99.5% of the uranium and plutonium are removed, including the same fraction of the other uranics, as well as significant amounts of the transuranics. Reprocessing, followed by concentration of the waste, reduces the waste volume by a factor of 20.[6] After 300 years, when the short-lived isotopes have all decayed, the total radioactivity of the waste is 10 times lower than that of unreprocessed fuel.[6]

But reprocessing is not all peaches and cream. Much high-level waste is produced. The treatment releases gaseous fission products to the atmosphere, and the fuel rods that have been chopped up in the process now contain neutron-activated radioisotopes. The initial liquid volume of the wastes is considerable. On the other side of the coin is the fact that reprocessing spent fuel dramatically lowers the need for the mining of new uranium, as well as the associated mining, milling, and enrichment. These pros and cons have led to much debate about the advisability of reprocessing. However, if one is convinced, as I am, that high-level wastes can be disposed of with great safety, then reprocessing makes very good sense.

Reprocessing is now being done in the United Kingdom, France, Germany, the USSR, India, and Japan, while five other countries contract with the United Kingdom or France for the reprocessing of their fuel. Nevertheless, only 5% of all nuclear fuel outside the communist world is reprocessed. Belgium and the United States have reprocessed in the past but no longer do so. In 1977, the Carter administration's "nonproliferation" policy indefinitely deferred the building of American civilian reprocessing plants or breeder reactors, and discouraged their development in other countries. The rationale was that such reprocessing produces purified plutonium, which could too easily be used for nuclear bombs. However, at the same time, we had the world's largest nuclear weapons arsenal, which in turn depended upon *military* reprocessing plants! It is clear that reprocessing, *combined with breeder reactor technology*, would give us unlimited fuel for electric power and would reduce the problems of high-level radwaste disposal.

It is important to bear in mind that the volume of nuclear-power-plant spent-fuel waste is incredibly small, compared with that of other power technologies. A *year's* waste from a large 1000-megawatt nuclear power plant, after reprocessing, weighs only 15 tons and—uranium being 19 times as dense as water— occupies a volume of only 2 cubic meters, which would fit nicely under a dining-room table. Put another way, one person's nuclear waste for a year's electricity can be reduced to the volume of a cigarette. In contrast, a 1000-megawatt coal-fired power plant uses enough coal *every day* to fill a train 100 cars long, and *every day* releases to the atmosphere 50 tons of fly ash (*after* electrostatic precipitation) and 500 tons of sulfur dioxide,[7] a primary cause of acid rain.

Dismantling

Nuclear power plants have a life expectancy of about 40 years. By that time, radiation damage to the structural components is so great that it would be hazardous to continue operation. The plant is shut down and the fuel rods are removed. However, a great deal of radioactivity remains, chiefly owing to the neutron-bombardment products in the steel reactor-core structures. Many of these induced radioisotopes are metals that were added in trace amounts to the steel in order to make it more resistant to corrosion ("stainless") or less brittle upon irradiation.

There are two important neutron-bombardment isotopes in these steel structures—nickel-59 and niobium-94—that have long half-lives and emit powerful gamma rays. Niobium-94 has the longer half-life (20,000 years), and its radioactivity will be more important than that of the nickel-59 after about 70 years. Anyone who stood close to such reactor components at the time of reactor shutdown would get a lethal dose of radiation in about 10 days.[8] Therefore, it is evident that some parts of decommissioned nuclear power plants cannot be classed as low-level waste, but must be put into a waste repository. This will be difficult and could cost 5–10% of the initial cost of plant construction.[8] A possible way out would be to develop steels with additives that produce less dangerous products upon neutron bombardment.

Waste Disposal: Time Factors

How long must a radioisotope be stored before it can be pronounced safe? *A rough rule of thumb is that an isotope becomes safe after storage for a time equal to 10 half-lives.* The radioactivity will then have dropped to $1/1000$ of its original level (see Figure 6).

Three hundred years of storage should therefore take care of anything with a half-life of 30 years or less. Such isotopes include all the common neutron-bombardment products, as well as some of the most common fission products, such as krypton-85, strontium-90, and cesium-137 (see Appendix C). But some of the transuranic radioisotopes, as well as neutron-activated isotopes like niobium-94, mentioned above, will require storage for much longer times.

The chief problem with high-level radwaste disposal is not the heat production, which is low after 40 years, or the gamma-ray emission, which may remain high. Both problems could be easily solved by erecting only minimal barriers, such as burial several meters underground. The major problem is the possibility of radioisotopes, especially the transuranics, getting into our food and water, so they could be ingested and remain in our bodies for a long time. Such ingestion exposes internal organs to the short-range alpha and beta particles, which are otherwise harmless if they remain outside the body.

Figure 24 shows the number of cancer deaths that would result if *all* of the reprocessed waste produced by one large

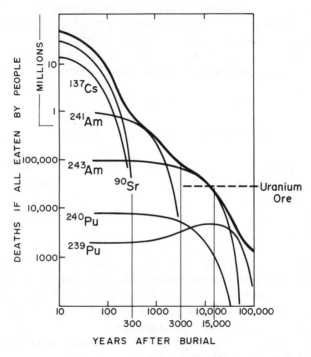

Figure 24. Ingestion hazard of radioactive waste. The graph shows (vertical scale) the number of cancer deaths that would result, if *all* the nuclear waste produced in 1 year by one large power plant, after reprocessing, were eaten by people, versus the number of years after burial of the waste (horizontal scale). The scales are logarithmic. The time scale begins at 10 years, after the waste has been held in a cooling pond. The heavy black curve—above all the others—shows the sum of the values of all the other curves. The broken horizontal line shows the radiation level of uranium ore. (Adapted from Figure 17 of Bernard L. Cohen, 1983, *Before It's Too Late*, Plenum Publishing Corp.)

nuclear power plant in 1 year got into food and water and was *all actually ingested* by people. This is clearly an extreme scenario! The radiation from the beta-particle-emitting fission products, such as cesium-137 and strontium-90, can be seen to drop rapidly, reaching low levels by 300 years (10 half-lives). At this time, the hazard from *all* the radioisotopes in the waste—the upper heavy curve in the figure—has dropped to $\frac{1}{100}$ of the original level. After 300 years, the hazard decreases more slowly, owing to the much longer half-lives of the transuranic isotopes americium-241 and americium-243, which emit alpha particles. Nevertheless, by 3000 years, the level is down to 3 times that of the originally mined uranium ore, and by 15,000 years is all the way down to the level of uranium ore. This level of waste, if all ingested by people in the U.S., would cause 30,000 deaths per year. But it is no more dangerous than the uranium ore currently present in the Earth. Obviously, people don't go around eating uranium ore. Nor would they ever ingest more than traces of nuclear waste, very little of which, after very deep burial, would ever reach the surface of the Earth.

Figure 24 also shows that the radioactivity from plutonium isotopes in reprocessed power-plant waste contributes relatively little to the total hazard. This of course would not be true of *un*reprocessed waste (spent fuel).

In summary, we have two quite different types of radwaste. *Most radwaste requires safe disposal for only about 300 years.* This includes most medical and industrial wastes, as well as most of the isotopes in dismantled nuclear power plants. The safe disposal of wastes for 300 years is not a great problem. One does not have to be concerned with such things as distant-future erosion or a major shifting of the Earth's crust. There is no problem with keeping records for that time. And, of course, as time goes on, the waste becomes progressively less dangerous.

But *reactor fuel waste requires long-term disposal*, whether it is spent fuel or reprocessing waste. We have seen that the radioactivity of reprocessing waste will drop to 3 times the level of uranium ore after 3000 years. Therefore, *reprocessing waste re-*

quires safe disposal for about 3000 years, chiefly because of the high concentration of the americium isotopes.

Unreprocessed (spent) fuel, on the other hand, will require 50,000 years to drop to 3 times the level of uranium ore, because it contains large amounts of long-lived transuranic isotopes.[6] Nevertheless, in spite of this slower decay, the United States plans to bury its spent fuel *without reprocessing,* because of largely unwarranted fears of the dangers of reprocessing technology. The Nuclear Regulatory Commission (NRC) and the Environmental Protection Agency (**EPA**) have decided that *spent fuel (unreprocessed) requires containment for up to 10,000 years,*[9] at which time the waste would be only about 10 times as active as natural uranium ore, and therefore relatively safe.

Long-Term Disposal of Waste[10]

We noted at the beginning of this chapter that *dispersal* is not a useful option for dealing with radwaste, except for low concentrations of radioisotopes with short half-lives. For example, tritium used in biological experimentation can usually be washed down the sink, because the amount is minute and dilution in the sewage system quickly reduces its concentration to natural levels. For high-level wastes, however, elaborate techniques of disposal are required. These techniques are still in the exploratory stage, but we must soon make decisions in this area, for the wastes are piling up.

Most high-level waste is reactor waste, which includes spent fuel, reprocessing waste, cooling-system filters, and parts of dismantled power plants. It may also include small amounts of industrial isotopes. The first step in handling reactor waste is usually *holding* it for anywhere from 10 to 40 years in on-site cooling ponds. Holding, of course, does not solve the long-term problem. For this, we may proceed to **confinement**, meaning long-term storage under conditions that are safe, but that permit retrieval of the waste by future generations. An alternative is

isolation of the waste, which does not permit future access. But in either case, the waste must first be concentrated, immobilized, and shielded, so that it can be safely handled and will be resistant to leaching by groundwater.

High-level waste can be *concentrated* by distilling it or by centrifuging it to remove water. It may also be treated with chemicals to render the waste insoluble. Next, it must be *immobilized*. Figure 25 shows how liquefied waste can be converted into a fine powder by exposing it to high temperature, which eliminates volatile substances. It may then be mixed with particles of glass (frit), poured into a stainless-steel canister, and melted to form a block of glass within the canister. In this way, the waste is solidified into a glass and surrounded by stainless steel. Forming it into a glass immobilizes it, rendering it resistant to erosion. Within the steel canister, it is *shielded*, so that it can be safely handled. The canisters will be buried in deep underground chambers. Such canisters are about 1 foot in diameter and 10 feet long. A year's waste from a 1000-megawatt reactor, after reprocessing, requires only 10 such canisters.

Confinement

The most secure mode of confinement is burial deep underground, called *geologic disposal*. The reason for such deep burial is not to shield ourselves from the gamma rays, which will penetrate no more than 10 feet through rock, but to provide a stable environment in which the waste will not get into groundwater in significant amounts for a very long time—ideally up to a million years. It must be recognized that, below about 200 feet, the earth is extremely stable and changes only very slowly over millions of years, except in regions with volcanic activity. Some parts of the continents—such as the Canadian Shield, centered in Ontario Province but extending into Minnesota and Wisconsin—are largely composed of granite, and have been stable for several *billions* of years. In deep underground confinement, waste canisters would be placed in trenches half a mile

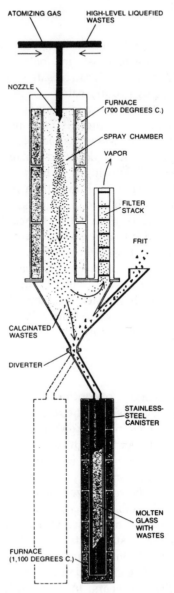

Figure 25. Treatment of concentrated high-level radwaste, resulting in both immobilization and shielding. (From Bernard L. Cohen, "The Disposal of Radioactive Wastes from Fission Reactors." Copyright 1977 by SCIENTIFIC AMERICAN, Inc. All rights reserved.)

underground, in very stable geological formations, such as granite, basalt, or salt.

The Nuclear Waste Policy Act, signed into law by President Reagan in January 1983, required the Department of Energy (**DOE**) to select and study five possible long-term geologic disposal sites, with the goal of having one go into operation by 1998. In May 1986, the DOE had narrowed the choice down to three sites, all west of the Mississippi: near Hanford, Washington (basalt), Amarillo, Texas (salt), and Las Vegas, Nevada (volcanic tuff).

The least satisfactory of the DOE site choices was the one at the Pasco Basin Cohasset flow, on the Columbia River near *Hanford*. The disposal site would be at least 3000 feet deep in miles-thick basalt, which is hardened volcanic lava. It appeared to be an excellent site. However, such basalt fields are often interspersed with loose layers of different rock that permit the passage of water—in short, aquifers. The site thus requires extensive test drilling to ensure that it will be resistant to water leaching. In addition, the repository would require the drilling of many 15-foot-diameter shafts 3000 feet deep through the basalt, a task beyond present capabilities.

Probably the next best site was the *Palo Duro Subbasin* salt, in the Texas Panhandle, southwest of Amarillo. The Ogallala aquifer, one of the largest in the U.S., and one that farmers in a wide area depend on for irrigation water, lies in this region at a depth of from 100 to 400 feet. The DOE proposed building the disposal site 2000 feet below the aquifer, from which it would be separated by at least 1000 feet of salt, as well as other stable rocks. Aside from the possible leakage of aquifer water into the vertical disposal shaft, a problem not hard to solve, the disposal site would be very stable. The Texas salt formations have not been invaded by groundwater for several hundred million years. Add to this the time it would take for any such water to leach the radwaste out of the glass and stainless steel in which it is enclosed—plus the fact that the radwaste, as it rises to the surface, is very effectively filtered by the rock it passes along the way—and it's not hard to see why the waste is unlikely to appear at the surface or in groundwater sooner than 1 million years after burial, and then only in the smallest amounts. Figure 26 shows the estimated maximum probabilities of releases from a 700-foot-deep clay repository. It can be seen that, after 100,000 years, there is less than one chance in a million—a probability of 10^{-6}—of radioactive release to the air, one chance in a thousand (10^{-3}) of release to the land surface, and one chance in a hundred (10^{-2}) of release to groundwater.

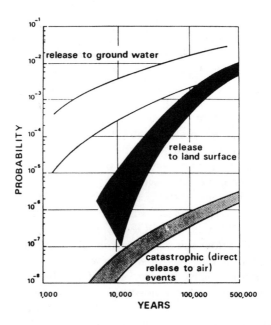

Figure 26. Ranges of maximum probabilities (within curves) for different scenarios of radwaste release from a model of a 700-foot-deep clay repository at Mol, Belgium. (Reprinted with permission from M. d'Alessandro and A. Bonne, in G. de Marsily and D. F. Merriam (Eds.), *Predictive Geology*. Copyright 1982, Pergamon Press PLC.)

But the most promising site was at *Yucca Mountain*, about 90 miles northwest of Las Vegas. This is a ridge some 1500 feet *above* the surrounding terrain, composed of welded volcanic tuff, a substance something like cinder block, but more dense. The water table is 1000 feet *below* the surrounding terrain. Waste stored within this mountain would be easily accessible from the sides and would always be above the water table. The annual rainfall is 6 inches, most of which evaporates. Flow in the water table is very slow, and moves toward Death Valley, some 50 miles west. The area is owned and guarded by the federal government. The nearest inhabitants are in the small town of Beatty, 15 miles away. Although underground tests of nuclear weapons are performed about 30 miles away at the Nevada Test Site, the Environmental Protection Agency has not observed notable amounts of fission products or tritium in wells only one-quarter mile away from the explosion sites, and tests show that vibration from the tests is not a problem. Furthermore, underground testing of nuclear weapons may well cease altogether before Yucca Mountain becomes a permanent facility.

The DOE had proposed extensive testing at all three sites, a very expensive and time-consuming process. However, the Gordian knot was severed with one bold stroke when Congress, in December 1987, directed the DOE to focus exclusively on the Yucca Mountain site, leaving the Texas and Washington sites in second and third place, respectively.[11] The State of Nevada would receive $20 million a year as "incentive money." To balance the decision geographically, a requirement of the original Nuclear Waste Policy Act, it was proposed that a Monitored Retrievable Storage (MRS) facility be built east of the Mississippi at Oak Ridge, Tennessee, to store spent fuel in a safe but retrievable way until the Nevada repository opens.

Many nontechnological factors enter the decision for choosing an ideal disposal site. They include avoidance of areas where (1) population density is high; (2) valuable ores might occur, leading to possible future drilling; or (3) extensive aquifers exist, which might be used by farmers or city populations. The Hanford and Palo Duro sites satisfy the first two criteria quite well but are weak on the third. Yucca Mountain, on the other hand, satisfies all three criteria and is much more promising than the other sites.

Nevertheless, tremendous opposition to all three sites has developed, chiefly by the people living in the regions concerned. Such opposition has usually been exacerbated by an alarmist press. The fire is fueled by local politicians, most of whom know very little about the disposal of radioactive wastes. Many are only too ready to play on public fears and will object to any kind of waste disposal in their districts, regardless of how well-designed these projects may be. This reaction is commonly known as "Not-In-My-Back-Yard," or NIMBY. Science journalist Luther Carter cited one Department of Energy official, who commented on the "unrelieved negativism" of Texas state officials, when faced with the prospect of a high-level waste repository in the Texas Panhandle:

> They have reinforced and confirmed the worst dreads expressed by any of the local opposition. . . With no holding back, very quickly a frenzy of demagoguery develops, much like the theatrics of domestic war propaganda except in this case DOE and nuclear waste are the evil.[12]

This is not to say that the salt beds of the Texas Panhandle would make an ideal high-level waste disposal site: they might not. But the DOE was simply proposing that test drills be made as part of *initial studies* for determining site feasibility. It would take a decade of studying this site under a microscope, so to speak, in order to be thoroughly assured of its safety before it is used. Yet land values in the area immediately fell because of public fears.

In a similar vein, opposition to the Yucca Mountain proposal is growing in Nevada, and it is now uncertain that the repository will be built. Extensive testing still needs to be done before Yucca Mountain can be the final choice, but the state has refused to issue permits for geological characterization of the site.

Finally, confinement achieved by dropping waste canisters into the ocean floor is currently being investigated. Many ocean sediments are quite soft. A streamlined canister, with stabilizing fins, that contains high-density waste fuel, would drive into this sediment like an arrow, sinking perhaps 100 feet into the silt. It would sit there for millions of years. It is estimated that plutonium-239 in such a site would not migrate more than a few meters from a *breached* canister in 100,000 years.[13] Even moderate leaching by seawater would result only in extremely slow leakage out of the silt and into the seawater, where it would immediately undergo tremendous dilution, so that it would be of no harm to marine life. This would be one of the cheapest and easiest ways to dispose of high-level radwaste. In addition, the waste would be accessible to future generations. However, unwarranted public fears that seabed burial would be unsafe totally preclude this option at the present time.

It is time that Americans came to realize that they have opted for nuclear power and nuclear weapons, and now they have to do something about disposal of the wastes. Both the Yucca Mountain repository and careful seabed burial show promise of being extremely safe options. The wastes exist and must be disposed of. They *can* be disposed of safely. We have already wasted too much time resisting rational disposal of high-level nuclear wastes. Those who fear and oppose long-term nuclear waste disposal have an obligation to let the rest of us know what they propose as an alternative.

Isolation

Many of the deep-disposal confinement schemes discussed above could be used to isolate the waste, simply by sealing off the site.

However, many isolation schemes are more exotic. For example, we could put the waste into a space rocket and shoot it into the Sun. The Sun is a tremendous nuclear reaction machine, and the addition of our puny amount of radwaste would never be noticed. This option would get rid of the waste forever! But there are two difficulties. First, the procedure is very expensive. Second, if there were an explosion upon launch of the rocket, the radioactive waste could be strewn about in a disastrous way.

Another exotic possibility is to place the waste canisters several hundred feet under the ocean floor in one of the deep-sea trenches that occur, for example, off the north coast of Puerto Rico and the east coast of Japan. These trenches are in what are called *subduction zones*, where the ocean floor is very slowly sliding under the continents and is being pushed down into the molten rock of the Earth's mantle. Waste buried in these trenches would eventually join all of the other radioactive material deep in the Earth that is responsible for the molten rock in the first place. Nuclear surface ships and submarines could be scuttled at such sites. The only problem at present with this scheme is that we don't know enough about the geology of such subduction zones. It is, however, an intriguing idea.

A third, more mundane, possibility is disposal in very deep boreholes, drilled down approximately *five miles*. After insertion of the waste and removal of the drilling pipe, the drill hole would seal up, because of the high pressure and temperature of the surrounding rock, aided by the heat of the radwaste itself. Radwaste buried this way would be securely isolated from groundwater for millions of years. The technique is expensive, however, and would probably be feasible only for the smaller volumes represented by reprocessing waste; it is being seriously considered by several countries.

As I have noted, these isolation schemes have both positive and negative aspects. On the one hand, the waste is gone forever, as far as the human race is concerned. On the other hand, future generations are deprived of access to the waste, which they might someday find quite valuable.

In summary, only high-level waste poses a long-term hazard. This waste consists mostly of spent fuel, but in the future, it will also include reprocessing wastes and parts of dismantled plants. This material can be put into safe disposal for a very long time, although such disposal will add to the total cost of nuclear power. Most of the danger of high-level radioactive waste is gone after 300 years, and it might be better simply to confine rather than isolate the waste for that time, thereby permitting future generations access to it. By that time, human society will presumably be more knowledgeable about radwaste and could decide either to dispose of it in a more permanent way or to exhume it and extract valuable radioisotopes.

10 ⚛ Myth I: Nuclear Power Is Too Dangerous

The general public is afraid of nuclear power. This technology seems to be a ferocious lion, raising feelings of dread. Much of this attitude results from a false perception of the dangers, a perception that arises largely from the mystery that surrounds nuclear energy, and from the conviction that nuclear radiation is deadly. As with our ideas of lions, there is, of course, an element of truth in these perceptions; the problem is that they are so highly exaggerated. Let us look a little more closely at our perceptions of danger.

One of the most fundamentally hazardous activities that humans engage in is flying in airplanes. Getting into a sheet metal tube and flying 7 miles above the Earth at 600 miles per hour, knowing that an accident causing decompression would kill everyone instantly, would appear to be sheer insanity. Yet we do it every day. What we have done is *to take an inherently unsafe activity and make it very safe*. Most of the aviation pioneers died flying. But gradually, over the decades, we made flying safer, until by 1940, even heads of state were willing to risk their lives in order to get to a meeting in another country much faster. Today, flying in a scheduled airliner is 30 times safer than riding in a car. Flying is now so remarkably safe that a major crash of a scheduled American airliner occurs only about once a year and is considered big news when it happens.[1]

Some people, of course, are still afraid of flying. Most of them recognize that their fear is psychological rather than based on a rational assessment of the dangers. "Fear of flying" has now become an expression that may mean fear of doing anything new, exciting, and possibly dangerous, such as getting married or taking a responsible new position. In all of these cases, the person's fear may be just as irrational as a fear of flying.

This is not to say that we don't understand such fears. No one laughs at a person who is afraid to look down from the top of the Empire State Building. We understand this, even though it may seem irrational. In fact, our understanding of such fears often enables us to teach people how to overcome them. One cure is simply to do the feared thing frequently. People who fly all the time usually become quite blasé about it. We even have an expression for this: *Familiarity breeds contempt.*

A prime example in our modern society is automobile driving. Hardly anyone is afraid of riding in a car. Yet that would indeed be a rational fear, as automobile travel really is dangerous. Imagine what a reaction you would get if you took an intelligent person who had never seen a car—like Benjamin Franklin—for a drive down a 2-lane state highway. Not only would you be moving at what your passenger would consider an incredible and extremely unsafe speed, but every few seconds you would miss a car coming in the opposite direction by only a few feet. Old Benjy would go bananas! His fear would be rational: a head-on collision of two of these cars would probably kill all the occupants. Yet we take this risk all the time, and think nothing of it. People think so little of it that they even resist wearing a seat belt. Seat belts are now quite comfortable, so this attitude is irrational. It reveals the extent to which our familiarity with the automobile has bred a contempt of its dangers.

Public Perception of Hazards

Figure 27 illustrates the fact that the public's perceptions of the dangers of certain activities may not be closely related to the actual hazard, as estimated by experts. Three groups of people were questioned: members of the League of Women Voters, college students, and members of business and professional clubs. The second column from the left shows the number of deaths per year from the activity—these are either actual deaths or, in some cases, like nuclear or electric power, the estimates of experts. For example, for civilian nuclear power, which has killed no one directly in the United States, the number reflects primarily the expected increase of lethal cancer due to radiation, among workers involved in uranium mining and fuel fabrication, as well as among the public as a result of estimated future reactor accidents. The total is only 100 deaths per year. If we compare this to the 500,000 cancer deaths per year from all causes, nuclear power would account for only $1/5000$ of all cancers. Natural background radiation accounts for about 2% of all cancers, or 100 times as many as nuclear power.

Another factor to be considered when interpreting these statistics is that some of the activities are performed by most of us, such as driving motor vehicles and using electric power, whereas some involve only a few of us, such as riding motorcycles and fire fighting. The latter are therefore more dangerous than their numbers would indicate: if as many people drove motorcycles as drive cars, the number killed by motorcycles would be much higher than the number killed by cars.

With these considerations in mind, let us look more closely at the results of this survey. The five greatest hazards, each accounting for more than 10,000 deaths per year, are smoking, alcohol, motor vehicles, handguns, and electric power. "Electric power" is used here to represent all electric power generation by fossil fuels (coal, gas, and oil), and it includes the hazards of coal mining and air pollution. Another nine hazards account for

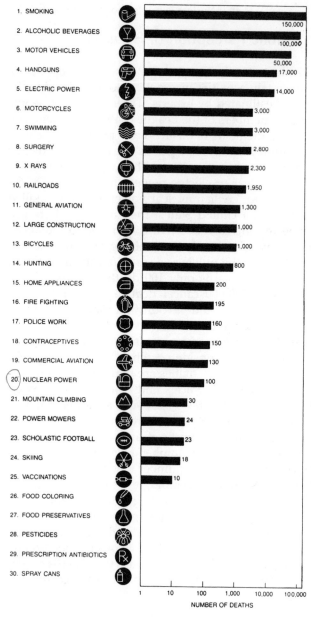

1. SMOKING		150,000
2. ALCOHOLIC BEVERAGES		100,000
3. MOTOR VEHICLES		50,000
4. HANDGUNS		17,000
5. ELECTRIC POWER		14,000
6. MOTORCYCLES		3,000
7. SWIMMING		3,000
8. SURGERY		2,800
9. X RAYS		2,300
10. RAILROADS		1,950
11. GENERAL AVIATION		1,300
12. LARGE CONSTRUCTION		1,000
13. BICYCLES		1,000
14. HUNTING		800
15. HOME APPLIANCES		200
16. FIRE FIGHTING		195
17. POLICE WORK		160
18. CONTRACEPTIVES		150
19. COMMERCIAL AVIATION		130
20. NUCLEAR POWER		100
21. MOUNTAIN CLIMBING		30
22. POWER MOWERS		24
23. SCHOLASTIC FOOTBALL		23
24. SKIING		18
25. VACCINATIONS		10
26. FOOD COLORING		
27. FOOD PRESERVATIVES		
28. PESTICIDES		
29. PRESCRIPTION ANTIBIOTICS		
30. SPRAY CANS		

NUMBER OF DEATHS

Figure 27. Data from a survey in which members of three groups were asked by Decision Research (1201 Oak Street, Eugene, OR 97401) to rank 30 sources of

LEAGUE OF WOMEN VOTERS	COLLEGE STUDENTS	BUSINESS AND PROFESSIONAL CLUB MEMBERS
NUCLEAR POWER	NUCLEAR POWER	HANDGUNS
MOTOR VEHICLES	HANDGUNS	MOTORCYCLES
HANDGUNS	SMOKING	MOTOR VEHICLES
SMOKING	PESTICIDES	SMOKING
MOTORCYCLES	MOTOR VEHICLES	ALCOHOLIC BEVERAGES
ALCOHOLIC BEVERAGES	MOTORCYCLES	FIRE FIGHTING
GENERAL AVIATION	ALCOHOLIC BEVERAGES	POLICE WORK
POLICE WORK	POLICE WORK	NUCLEAR POWER
PESTICIDES	CONTRACEPTIVES	SURGERY
SURGERY	FIRE FIGHTING	HUNTING
FIRE FIGHTING	SURGERY	GENERAL AVIATION
LARGE CONSTRUCTION	FOOD PRESERVATIVES	MOUNTAIN CLIMBING
HUNTING	SPRAY CANS	LARGE CONSTRUCTION
SPRAY CANS	LARGE CONSTRUCTION	BICYCLES
MOUNTAIN CLIMBING	GENERAL AVIATION	PESTICIDES
BICYCLES	COMMERCIAL AVIATION	SKIING
COMMERCIAL AVIATION	X RAYS	SWIMMING
ELECTRIC POWER	HUNTING	COMMERCIAL AVIATION
SWIMMING	ELECTRIC POWER	ELECTRIC POWER
CONTRACEPTIVES	FOOD COLORING	RAILROADS
SKIING	PRESCRIPTION ANTIBIOTICS	SCHOLASTIC FOOTBALL
X RAYS	MOUNTAIN CLIMBING	CONTRACEPTIVES
SCHOLASTIC FOOTBALL	RAILROADS	SPRAY CANS
RAILROADS	BICYCLES	X RAYS
FOOD PRESERVATIVES	SKIING	POWER MOWERS
FOOD COLORING	SCHOLASTIC FOOTBALL	PRESCRIPTION ANTIBIOTICS
POWER MOWERS	HOME APPLIANCES	HOME APPLIANCES
PRESCRIPTION ANTIBIOTICS	POWER MOWERS	FOOD PRESERVATIVES
HOME APPLIANCES	VACCINATIONS	VACCINATIONS
VACCINATIONS	SWIMMING	FOOD COLORING

risk. (From Arthur C. Upton, "The Biological Effects of Low-level Ionizing Radiation." Copyright 1982 by SCIENTIFIC AMERICAN, Inc. All rights reserved.)

deaths in the range of 800 to 3000 per year. The remaining hazards are relatively harmless and cause fewer than 200 deaths per year: they include commercial aviation, nuclear power (Number 20), food preservatives, and pesticides.

Now let's look at the *public perception* of these risks. Nuclear power is Number 1 on the list for both college students and members of the League of Women Voters. It is Number 8 for business and professional club members, which suggests that these people tend to estimate risk more as the experts do. Note that all three groups of people included motor vehicles and smoking among the top five, where they belong; this suggests that the public's perception of the dangers of these activities is quite accurate. That the public accurately assesses the danger of motor vehicles may seem to contradict my earlier statement that people do not worry about automobile accidents. In fact, both statements are true. The public is very well aware that roughly 50,000 people die every year on our streets and highways, but their familiarity with the automobile blinds them to the possibility that they may themselves become one of these 50,000 casualties.

For several items, the hazards were consistently overestimated or consistently underestimated by *all* groups. The dangers of pesticides were greatly overrated, and those of home appliances were greatly underrated. This is another example of familiarity blinding us to real dangers. The consistent underestimate of the dangers of X rays is a remarkable example of a technology that has become accepted because we are used to it, even though X rays are invisible and, to most people, mysterious.

The survey reveals one particularly unexpected result: the public estimates the dangers of both general and commercial aviation quite realistically. How does this estimation jibe with our discussion of "fear of flying"? One explanation could be that the groups interrogated represented fairly cosmopolitan individuals, who were likely to do a fair amount of flying and, because of their familiarity with it, were not afraid. It is interesting that the public, while apparently *understanding the dangers* of both flying and driving, has a disproportionately low *fear* of driving and a disproportionately high fear of flying.

These psychological responses have been studied in a new area of research called *risk analysis*. These studies reveal that people's *perception* of the risk of an activity rises in proportion to their:

1. *Unfamiliarity* with it.
2. *Dread* of it. This includes what is called "catastrophic potential": once-a-year airline crashes are dreaded, although they kill no more *per year* than the 100+ people killed *daily* by automobiles.
3. *Lack of control* over it: *I* drive the car; *someone else*, whom I cannot control, flies the airplane.
4. *Lack of voluntary choice* associated with it: *I* decide to drive; nuclear power plants are *imposed* on me.[2]

It is apparent that the lay public has a much more complex definition of risk than the experts do. The experts assess risk numerically, or statistically, whereas the public assesses risk emotionally. *Although this may be irrational, it is not stupid.* It is not unreasonable to fear some kinds of death more than others. Most of us, regardless of the numerical risk associated with the activity, have a quite understandable fear of drowning at sea, or of a long descent to the ground in a doomed airliner. We have little fear of our death in an auto collision—even though it is far more likely—because we are very used to auto travel, and we have some personal control over the outcome. Full understanding of the risk of an activity may not even allay the fears of the experts themselves. This phenomenon affects newspaper reporters, too, who often magnify public misperceptions by their own emotional reactions to technologies that they either fear or do not fully understand.

Lack of voluntary choice is a strong component of public perception of risk. The person who is very nervous about a nuclear power plant being built near her or his home, or a nuclear waste disposal site 30 miles away, thinks nothing of swimming in a fast stream, riding a bicycle, or hunting—all rather dangerous activities. Yet the nuclear power plant not only is less hazardous, but probably contributes more to her or his general happiness than these sporting activities.

Such subjective factors lead to the result that, *in terms of lethality, the danger of nuclear power is greatly overestimated by the layperson, while that of fossil-fuel power is greatly underestimated.*

These insights into the perception of risk go far toward explaining public attitudes about nuclear power. Like commercial aviation, nuclear power does indeed have potentially dangerous aspects. But, also as in commercial aviation, the technology is remarkably safe.

Safe Aspects of Nuclear Power

We now review some aspects of nuclear power that are very safe, compared to other technologies; these were discussed in some detail in Chapter 7.

Normally operating nuclear power plants do not contaminate the atmosphere. Under normal operations, there is only a minute release of radioactive materials to the air from a nuclear power plant. The United Nations Scientific Committee on the Effects of Atomic Radiation reported in 1972 that the radiation doses at the boundary fences of 24 nuclear power stations in seven countries were well within the International Commission on Radiological Protection (ICRP) limit of 0.5 rem per year for the general public. The 22 modern versions of these nuclear stations gave doses of only about 0.005 rem per year,[3] which is the standard now accepted in the U.S. for exposures at nuclear plant boundary fences. This standard is only 1/60 of the total natural background radiation dose that we receive. Needless to say, no one lives at the boundary fence of a nuclear plant, and the radiation exposures drop rapidly at greater distances. The British have been unable to detect any iodine-131 at all in the milk of cows on farms near their Magnox nuclear reactors, which have been operating for decades.[3] Unlike fossil-fuel plants, nuclear plants release no noxious gases that will alter the climate or harm living organisms, nor do they make any soot or smoke that will despoil the countryside. In short, a normally functioning nuclear power plant is a very clean operation.

The danger of transporting nuclear fuel and nuclear wastes is very small. Nuclear materials are transported in exceedingly safe packages. No accidents anywhere in the world involving the transport of nuclear materials have resulted in harm to humans from released radioactive material.

Finally, *there has been no sabotage of nuclear power plants or materials.* Sabotage is, like the crash of an airplane, a possible but unlikely event. The safeguards that were established in the early days of nuclear power and nuclear weapons have proved to be effective.

Potentially Unsafe Aspects of Nuclear Power

There are some aspects of nuclear power technology that are a cause of reasonable concern. These problems generally have satisfactory technical solutions, but political difficulties may prevent their implementation. We discuss the three most important areas of concern.

A Reactor Accident with a Large Release of Radioactivity

With the exception of Chernobyl, only one accident—at the Windscale military reactor in England—has released significant amounts of radioactivity to the environment. The biggest American civilian accident was Three Mile Island, where 15 curies of iodine-131 was released to the nearby population, producing an exposure equivalent to two weeks of background radiation. The only lethal reactor accident in the U.S. was at the Idaho National

Engineering Laboratory in 1961, when a military reactor vessel exploded and killed 3 people.

The Chernobyl accident, of course, involved a very large release of radioactivity. But there are several important factors to consider: (1) the accident was the result of an unusual experiment, and not of normal operation of the reactor; (2) had the Chernobyl reactor been like most of those used elsewhere in the world, the accident almost certainly would not have happened; and (3) because it is impossible for a nuclear reactor to explode like a nuclear bomb, Chernobyl was about the worst reactor accident that could have happened. *These considerations make it unlikely that we shall ever see another Chernobyl.* Remember that most meltdowns do not involve a large radioactivity release to the environment. Furthermore, current reactors will gradually be replaced by inherently safe reactors, which are virtually meltdown-proof (Chapter 8).

Inadequate Waste Disposal

Waste disposal is a reasonable concern, especially because of the long half-lives of some nuclear wastes. However, the problems have been exaggerated. Science journalist Luther Carter noted:

> John Holdren, a physicist and energy resources specialist at the University of California at Berkeley, is an academic of high standing who is widely respected as a responsible and knowledgeable nuclear critic. Holdren recognizes that bringing about safe terminal disposal of radioactive waste is a significant problem, but, in relative terms, he ranks it last among the major problems of nuclear power. . .
>
> I was to find that Holdren's ranking of the nuclear risks is concurred in by a number of other scientific critics or opponents of nuclear power, such as Amory Lovins of Friends of the Earth, Thomas B. Cochran of the Natural Resources Defense Council (NRDC), and Terry R. Lash, formerly of NRDC and now director of the Illinois Department of Nuclear Safety.[4]

There is little doubt that we *know how* to dispose of radioactive waste in a safe way. The major problem is simply *making sure that it is done*. We have outlined in the previous chapter how the best efforts of the Department of Energy to establish long-term nuclear waste repositories have met with vigorous opposition. Politics, economics, and inaccurate public perceptions are major obstacles that must be overcome.

Proliferation

Some people fear that the **proliferation** of nuclear reactors around the world for the generation of electric power may enable more countries to build nuclear weapons. Is this a reasonable concern?

The resistance to the proliferation of nuclear technology arises because most reactors make plutonium. Plutonium is of primary concern because it makes a more efficient bomb than uranium. In addition, because plutonium is *chemically* different from uranium, it is easier to separate plutonium isotopes from uranium isotopes (in the spent fuel from a reactor) than it is to separate uranium isotopes from one another. To make uranium bombs, U-235 has to be purified (separated from U-238) to a concentration of about 90% from the 3% purity it has in reactor fuel. This is very difficult to do.

However, in a typical power reactor, the fuel is left in the reactor for three years, after which about 30% of the weapons-useful plutonium-239 has been converted to Pu-240. The Pu-240 emits neutrons profusely, which could cause a bomb to ignite prematurely, or "fizzle," instead of exploding. Therefore, the spent fuel from a typical power reactor is not very useful for making weapons. Weapons-grade plutonium is much more efficiently produced in a **plutonium production reactor**. The fuel is left in such a reactor for only about 30 days, which results in much purer Pu-239. This is how the military makes material for nuclear bombs. In both cases, however, the plutonium must be separated from the uranium in a reprocessing plant and purified to a level that is suitable for weapons.

Breeder reactors, discussed in Chapter 7, are highly efficient in making (breeding) plutonium from U-238. Therefore, the spread of this technology to many nations is a matter of concern, and it led the Carter administration to oppose further development of breeder reactors. However, extraction of plutonium from the U-238 blanket of a breeder reactor is in principle the same as its extraction from the uranium in the fuel rods of a power reactor or a plutonium production reactor. So, the important necessary link between reactors and bombs is the fuel-reprocessing plant. *The problem of proliferation therefore hinges on the siting of reprocessing plants* rather than on the siting of power, production, or breeder reactors. It should be possible, by international agreement, to have only a small number of heavily safeguarded reprocessing plants in the world. These could, for example, be located only in the countries which now have nuclear weapons (the United States, the USSR, Britain, France, and China). These countries could do the reprocessing for all other countries, and could thus control world plutonium stocks.

The *underdeveloped* country that is given a nuclear power reactor and is supplied with the necessary fuel cannot easily make a nuclear bomb. Reprocessing technology is difficult and expensive. To make one's own plutonium and to use it to make a bomb requires advanced technology and a cadre of skilled mechanics, engineers, and scientists.

It is the countries of *intermediate* technological development that cause concern. India is a good example. In the early 1970s, the Canadian government sold a heavy-water reactor to India, with the stipulation that it was to be used only to produce energy. The Indians, however, elected to operate the reactor in a mode that would produce weapons-grade plutonium. They purified this plutonium and proceeded to conduct a demonstration nuclear explosion in 1974. Then a funny thing happened. They announced that they now had the capability of making a militarily useful bomb, but that *they had no plans to do so*. Apparently, they wished to prove to themselves that they could make a bomb if they wanted to. Having done this, they were either satisfied with their own abilities, or they had some trepidation about becoming a nuclear-weapons nation. What seems to be most likely is that they now know that they can make a bomb fairly quickly, thus allaying their fears of possible aggression from Pakistan, a nation that probably has nuclear bomb capability. It should be borne in mind that India and Pakistan have many well-trained physicists and engineers—their successes could not be easily duplicated by an underdeveloped country.

It is of the greatest interest that, while about 30 nations now have nuclear power reactors, only 6 have demonstrated that they have nuclear-weapons capability. These include the 5 noted above, plus India. No other countries have produced nuclear bomb explosions since India's demonstration explosion almost two decades ago, although South Africa, Israel, and Pakistan undoubtedly have this capability. This situation strongly suggests that rational nations of intermediate development see no real gain in the long run in developing nuclear weapons.

Of course, we remain concerned about possibly irrational countries. But it is hard to imagine that a small country would dare attack one of the nuclear powers with a nuclear weapon. If the weapon were in a long-range missile, satellites would show precisely where it had come from, and the result would be instant and devastating retaliation. Such an attack would also invite disaster if it were directed against another small country that had a defense agreement with a nuclear power. Finally, holding nuclear weapons as a *deterrent* is probably effective only for a major world power.

Nevertheless, nuclear proliferation is indeed a rational concern, and it may represent the only truly rational argument against nuclear power. The world at large has been well aware of this. Ever since 1956, the International Atomic Energy Agency (IAEA) has monitored the use of nuclear fuel throughout the world and has applied safeguards against the diversion of fuel or reactors from civilian to military uses. The Nuclear Non-Proliferation Treaty of 1970 has been signed by 134 countries. China and France, unfortunately, have not yet signed but have stated their sympathy with its provisions. These are that (1) nuclear-weapon states shall not transfer weapons or weapons technology to non-nuclear-weapon states; (2) non-nuclear-weapon states will not attempt to acquire nuclear weapons; and (3) all parties will accept the controls and safeguards of the IAEA. As a result of this high level of awareness, the situation seems currently to be under control, but it bears watching. Clearly, we should press for *all* nations of the world to sign the Non-Proliferation Treaty.

Although nuclear proliferation can be used as an argument against the spread of nuclear power to nations that do not now have it, the prospect of conversion from civilian to military uses does not apply to the nuclear-weapon powers. Therefore, this is *not* a valid argument against the continued development of nuclear power in the U.S., except insofar as the development of plutonium breeder plants might tempt us to sell the plutonium to non-nuclear countries.

In summary, in the three areas of concern—reactor accidents, waste disposal, and proliferation—where one might indeed have rational fears, it would appear that proper controls and careful management can make the situation acceptably safe. I would draw once again upon the parallel with commercial aviation: *an inherently unsafe technology can be made very safe if the will exists to make it so.*

Regulatory and Political Problems: Civilian

We are primarily concerned in this book with the technological promise of nuclear power and not with political considerations. Nevertheless, the politics of nuclear power is pervasive and cannot be ignored.

Plant Construction

Our current method of building nuclear power plants leaves much to be desired. Construction companies, used to building structures like bridges, where there are large safety factors to cover up occasional errors, have often built nuclear plants in a shoddy way. As a result, much construction has had to be ripped out and done over. It takes 13 years to build a typical nuclear plant in the U.S., from original design to start-up, whereas in France it takes only 6 years, and costs half as much money. These long construction times are not entirely the fault of the construction companies; they also reflect the often poorly coordinated and complex requirements of the Nuclear Regulatory Commission. Of course, we want safe plants, but a line has to be drawn somewhere; current bureaucratic red tape is excessive. The Comanche Peak power plant southwest of Dallas-Fort Worth, originally projected to cost $779 million, wound up costing $9.1 *billion*—almost a 12-fold increase. These excessive costs are passed on to the consumer and are one reason why U.S. nuclear power is often more expensive than fossil-fuel power.

Plant Siting

The siting of nuclear plants has often been surprising, if not alarming. A plant in Diablo Canyon, California, was built near an earthquake fault. Several reactors have been built too close, not just to cities, but to metropolises, such as the Indian Point nuclear site on the Hudson River, which is only 30 miles from Manhattan. Several have been built *upwind* of metropolises, such as Comanche Peak, about 50 miles southwest of the Dallas-Fort Worth metroplex.[5]

Waste Disposal

We described in the last chapter the general paranoia that has surrounded this issue, exemplified by the NIMBY reaction. There seems to be a lack of recognition by the public that the scandalous disposal practices at our military production reactors (see below) have *not* been followed by the civilian nuclear power industry, which has an excellent record on waste handling.

Public Perceptions

The failure of the nuclear reactor industry to build plants economically and to site them well away from population centers, as well as the failure of both politicians and the federal government to establish a sound policy for the disposal of either low-level *or* high-level nuclear waste, has contributed to the American public's distrust of nuclear power. Added to the unwarranted fright produced by the Three Mile Island accident, a general attitude of opposition to all nuclear power has arisen. As a consequence, there have been no orders for new nuclear power plants in this country for 15 years, and many existing orders were canceled during this period.

Many of the public's misconceptions stem directly from hysterical news stories. For example, in May 1986, the *New York Post* wrote that "as many as 15,000 people are already dead [and] being buried in a mass grave 150 miles from Chernobyl." The *Philadelphia Evening Bulletin*, reporting on Three Mile Island, proclaimed, "It's Spilling All Over the U.S." With this kind of sensational reporting, it's no surprise that people do not know that the *maximum* dose outside the Three Mile Island plant was less than one-tenth of a rem.

An example of a truly irrational reaction to nuclear power is the decision to shut down the Shoreham Nuclear Power Station on the north shore of Long Island. This plant was completed in 1985, after the expenditure of $5.3 billion on its construction. But it never went into operation. In May 1988, the New York State government took over the plant, with the intention of dismantling it. The plant faced fierce local opposition due to fears about safe evacuation of the surrounding population in the event of an emergency. Yet, this plant is (1) by the seashore, so that the probability of fallout over the land during an accident is cut in half; (2) located where the predominant winds are *offshore*; and (3) 50 miles *downwind* of New York City. In addition, it is equipped with a tank of boron water that could be pumped into the reactor core during an emergency, which would stop the fission reaction immediately. It would have been one of the safest plants in the U.S. Joseph W. McConnell, a vice-president of the Long Island Lighting Company, which owned the plant, made the sad comment, "In the politics on Long Island, it's no longer a debatable issue. People couldn't win on the rational side."[6]

But there are rays of hope. In July 1988, the Citizens Association for Sound Energy, a group that for years had objected to licensing of the Comanche Peak nuclear power plant in north Texas, reached a sweeping agreement with the operators, Texas Utilities Electric, and the Nuclear Regulatory Commission (NRC). This citizens' group accepted licensing of the plant and, in the process, acquired unprecedented leeway to oversee plant operations: they may monitor plant developments for at least five years after the start of plant operations, may attend previously closed meetings between the utility and the NRC, and may have access to construction sites and to correspondence with the NRC. The plant was licensed and went into operation in 1990. This initiative could be a model for future responsible involvement of citizens in nuclear power planning.

Regulatory and Political Problems: Military

The attitude has been: "The Russians are coming! Produce! Don't tell the public about the dangers. Dump the waste in the pits out back. We'll worry about them later."

SENATOR JOHN GLENN
Issues in Science & Technology, Summer 1989

Waste Disposal

In the 1960s and 1970s, the Atomic Energy Commission (AEC) permitted shocking practices in the area of military radioactive waste disposal.[4] At the Hanford facility in Washington State, which produced plutonium for nuclear weapons, high-level nuclear waste was placed in underground tanks that have since leaked, and solid wastes containing plutonium were simply buried in trenches. Cleanup may cost as much as $50 *billion*. The Savannah River plant in South Carolina, which produced plutonium and tritium for weapons, and which reprocessed nuclear fuel, also has serious waste problems, which may cost $10 billion to clean up. Similar problems have emerged at other nuclear weapons sites, such as those at Rocky Flats, Colorado, and Fernald, Ohio. Finally, waste disposal problems exist at the nation's national laboratories, the worst—in order of decreasing cost—being at the Idaho Engineering Lab; the Oak Ridge, Tennessee, lab; the Los Alamos, New Mexico, lab; and the Lawrence Livermore, California, lab. Cleaning up all the national laboratories will cost about $15 billion. The total cost of all these cleanups may come to $110 billion.[7] As Senator John Glenn notes, the chief contributing factor to this problem was the paranoid zeal with which we pursued our nuclear weapons program, a zeal that left little room for rational thought about responsible waste disposal. In the three decades after World War II, the AEC operated under a cloak of military secrecy, and few people knew of the extent of these transgressions. The situation has improved greatly since the Department of Energy (DOE) took over the control of military nuclear matters from the AEC in 1975, but only now are these problems becoming publicly recognized.

It must be recognized, however, that the waste-disposal hazards at the nuclear weapons plants will be removed. Furthermore, most of the waste is no longer accumulating. The plutonium and uranium used for bombs have very long half-lives: they last essentially forever and can be recycled from old weapons into new ones. Consequently, the DOE has not produced new plutonium in military reactors since 1988; and it ended enriched-uranium production for weapons in 1964.

In the meantime, studies show that military nuclear activities appear not to have endangered the people working at the sites. The death rates among nearly 10,000 white males who worked at the Savannah River plant in the period 1952–1974—a group we would expect to have received high exposures from poor waste-disposal practices—were found to be *lower* than the death rates in the general U.S. population (Figure 28). A subgroup of 1274 workers who were hired before 1955 and who worked 5 to 15 years at the plant did show a statistically significant increase in death from leukemia, but the numbers were only 6 deaths compared to 2.18 expected. This is the only statistically significant increase in deaths from any cause found in this study, conducted by the Center for Epidemiologic Research of Oak Ridge Associated Universities. Another study of workers at Hanford, Sellafield (England), Oak Ridge, and many other places showed that there is "no proof of an increased mortality

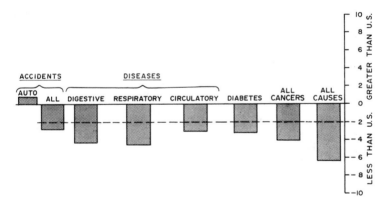

Figure 28. Death rates among 9860 white males who worked at the Savannah River, South Carolina, nuclear weapons plant in the period 1952–1974, compared to death rates in the general U.S. population. Numbers at the right are Freeman-Tukey residuals calculated from the observed and the expected number of deaths. Values greater or less than 2 are statistically significant. Only data for hourly workers are shown; the data for salaried workers were quite similar. (Redrawn from graph by Center for Epidemiologic Research, Oak Ridge Associated Universities, *42nd Annual Report, 1988.* Work was performed under contract number DE-AC05–76OR00033 between the U.S. Department of Energy and Oak Ridge Associated Universities.)

for workers in nuclear industry, either for all causes of death or for cancer or leukemia."[8] "No proof" does not exclude the possibility of some harm, but the harm would have to be small if it is so difficult to measure.

Weapons Plant Emissions

In the early days of nuclear weapons production, several unusual releases of radioactive gases from weapons reactors were not reported by the AEC. The worst was a release at Hanford, where 400,000 curies of radioisotopes leaked into the atmosphere from fuel-reprocessing tanks, which were not connected to vent filters in the early years of 1944–1947. Some 14,000 people may have received iodine-131 thyroid doses greater than 33 rads, and about 1400 children could have got thyroid doses from 15 to 650 rads.[9] For a few infants, total doses were as high as 2900 rads. These, of course, are doses to the thyroid and not to the whole body. A dose of about 30 rads is received in a diagnostic nuclear scan of the thyroid, and will probably do no harm. But the higher doses that some children received carried a significant risk of inducing thyroid cancer. This all happened a long time ago, and all the cancers should have developed by now, but it is difficult to track down the exposed people. In any event, those who got thyroid cancer would probably have been treated, and most of them cured. But this presumed favorable outcome does not relieve the AEC and the military of responsibility for such careless behavior.

Fallout from Tests

Another historic problem was the generally cavalier approach of the AEC and the U.S. armed services to weapons testing. In the 1950s and 1960s, members of the armed services were exposed to radiation from the atmospheric testing of nuclear bombs in the South Pacific and in the Nevada desert. It is difficult to understand why they so exposed these people, for any radiation biologist would have understood the hazard, even in those early days, and could have explained it to the AEC and the military. Although it has been found that in only a few tests was this exposure sufficient to cause cancer, the federal agencies were not candid at the time in their reporting of these events. The result has been strong suspicions by veterans and the public that these tests were extremely dangerous.[10] These actions have not helped the cause of nuclear power. Although nuclear weapons and nuclear power are quite different things, the AEC was involved in both, and questionable performance in one area inevitably produces doubts about the other area.

Regulatory and Political Problems: Overview

What are we to make of all this?

First of all, it is clear that the AEC and the military handled nuclear waste in a quite irresponsible way during the Cold War frenzy to build a nuclear arsenal 10 times greater than was needed. This was done under high secrecy, so few people knew about it. But civilian nuclear power is a totally different thing and has, to date, an excellent record on waste handling. It will cost at least $100 billion to clean up the weapons production mess, but this is not a reason to phase out *civilian* nuclear power.

The civilian nuclear power industry is certainly not perfect. There are still many inefficiencies, but these will be taken care of by the economic pressure of competition from other sources of power. There are some safety problems, such as poor plant siting, the faulty construction of some plants, and inadequate operator training. These are worth worrying about, but present public awareness, plus the fact that the Nuclear Regulatory Commission is a quite different beast from the old AEC, makes it most likely that such problems will gradually be solved. The next generation of nuclear reactors will be easier to build and much safer to operate.

In spite of all the problems, many of which are merely *per-*

ceived to be problems, the fact remains that *no one has ever been killed by civilian nuclear power in the U.S.*, and few people are likely even to have been harmed by it. This conclusion is underscored by a recommendation in 1989 from the Council on Scientific Affairs of the American Medical Association (AMA) that "generating electricity with nuclear power is acceptably safe in the United States."[11] The safety record of the civilian nuclear-power industry is almost incredible and quite without parallel in modern times.

11 ⚛ The Power Problem

The public must recognize that a risk-free society is not only impossible, but intolerably expensive. . . There are numerous deaths from falls down stairs in the home every year, but we do not advocate that all staircases be replaced by elevators.

DANIEL E. KOSHLAND, JR.
Editor, *Science* magazine
Science **244**, 7 April 1989

In any assessment of the desirability of sources of electrical power it is important to be aware of a central concept in risk analysis: *risk versus benefit*. Everything we do involves some risk. Driving to the grocery store is a risky activity. You cannot even avoid risk by staying in your house: even if you remain in bed and are waited on, your muscles will atrophy, and you will become ill simply as a result of inaction. Furthermore, you will still be subject to the dangers of fire, tornado, and so on. It would appear that the only perfectly safe thing to do is die!

We are willing to accept greater risks if the benefit will be greater. For example, we undergo the risk of surgery because it may cure us. We accept the high death toll of driving cars because the automobile is so useful to us in our modern society.

But people are sometimes irrational about weighing risk against benefit. For example, the use of an automobile seat belt involves no risk and very little discomfort, and the benefits are great. Yet, many people seem to have erected a mental block against using them; some people apparently feel that laws requiring the use of seat belts are an infringement on their freedom. Similarly, people may become very emotional, and even irrational, about sources of power. They tend to be quite accepting of power sources that have been in use for a long time, like coal, but to be very nervous about new sources, like nuclear fission. One way to cut through a great deal of the confusion about sources of power is always to keep in mind the risk versus the benefit.

The Power Dilemma

Let us look at some facts about power production. The kind of power we are considering is power that is used by a large segment of the population—primarily electric power. We are not concerned with individual or isolated sources of power, such as flashlight batteries or firewood.

We require two kinds of power: *concentrated* and *dispersed*. Concentrated power is needed for running factories and providing electricity for cities. Such endeavors require continuous day-and-night releases of large amounts of energy in a relatively small space. Dispersed energy involves small energy releases, usually on an irregular schedule, and spread across a larger space—as in running our cars, powering portable radios, or heating homes. The importance of this distinction is that *only certain energy sources provide concentrated power*.

Consider, for example, solar energy. If you live in a sunny climate and the weather is not too cold, then you can quite reasonably heat your home with solar power. You may even be able to provide enough power to heat your water and run your TV set. You could also heat a one-story school building with solar power; this use is particularly appropriate because schools don't have to be heated at night. Solar power *does* have a place in the power picture, and we should continue to develop it for these purposes. But you can't use solar power to run a big factory, like an aluminum refinery, which uses a great deal of electric power, because solar power is not sufficiently concentrated. And unless the energy could be efficiently stored, the factory could not operate in bad weather or at night. Frederick Seitz, former head of the National Academy of Sciences, commented:

> To meet our electrical needs, we'd have to build enough collector plates to cover the state of Delaware. No serious student of solar power expects it to be anything but a supplement to conventional electricity for decades.[1]

Similar arguments apply to power sources derived from solar energy, such as wind, waves, and falling water from rivers and lakes. Power from such sources is available at sufficient intensity in only a limited number of places. We have already exploited virtually all the available sites in the U.S. where falling water can give us hydroelectric power.

How about geothermal power: using the heat in the ground far beneath us? Again, in most cases, it is not sufficiently concentrated. It's great for heating a community of homes, but it cannot heat a large city that is far from the source of heat. Geothermal power is easily available in only a few places, mostly in California, Nevada, and Oregon—unless one plans to drill miles down into the Earth, which would be very expensive. (An interesting and not generally recognized fact is that geothermal energy is basically *nuclear* energy, produced by the radioactivity of the Earth!)

There is a great psychological appeal in using what people consider *limitless and nonpolluting* sources of power, like the wind or the sun. Although it is true that these sources are limitless, it is not true that they don't produce pollution. For solar power, you need solar collectors. The manufacture of these structures may produce pollution. When the collectors are worn out, they will be discarded—another source of pollution. And so on. There is no free lunch! Although solar and other limitless power sources are very attractive in many respects, they do have their costs.

There are only two feasible sources of concentrated energy: fossil fuels and fissionable or fusible atoms. **Fossil fuels** come from ancient sediments, over 100 million years old, and include coal, oil, and natural gas. Fissionable atoms are uranium and plutonium. Fusible atoms are hydrogen isotopes. Let us consider the pros and cons of these energy sources.

Fossil Fuels

Coal, oil, and natural gas are concentrated sources of energy that are relatively easy to mine and transport. When burned, coal and oil are dirty, coal being the worst offender, producing annually millions of tons of sulfur and nitrogen oxides and tens of thousands of tons of particles. The sulfur and nitrogen oxides cause *acid rain*, which kills plant life and fish in streams, lakes, and even coastal estuaries, such as Chesapeake Bay. The nitrogen oxides contribute to *ozone depletion*, which is a decrease in the concentration of the ozone layer in the upper atmosphere that protects us from dangerous ultraviolet radiation from the Sun. Finally, the carbon dioxide (CO_2) emissions from fossil-fuel plants are leading to a *greenhouse effect*, in which the CO_2 reflects heat back to the Earth, eventually raising the temperature of the atmosphere. The greenhouse effect could produce devastating alterations of agricultural patterns, as well as melting the polar ice sheets and raising water levels until coastal cities like Galveston and Miami are inundated. The gas and particle emissions from these coal-fired plants also cause respiratory problems[2] that produce lung cancer and that shorten lives. Coal smoke even contains radioactivity—as much as or more than is emitted by a nuclear plant (Table 2). Some of this pollution can be avoided by more efficient combustion of the coal, and by cleaning, or "scrubbing," the coal smoke. But these measures cost money and raise the price of the electric power. Nothing can be done about the CO_2.

It is clear that fossil-fuel technology is dangerous to the public at large. But David Sheridan, editor of the Ford Foundation's Energy Policy Project, explained how extremely dangerous it is to coal miners:

> In this century alone in the United States, more than 100,000 men lost their lives in coal mines. More than a million more were permanently disabled in mine accidents. Even more—no one knows how many—contracted pneumoconiosis (*black lung disease*) and spent their last years gasping for breath.[2]

Coal mining is safer now, but the industry has been slow in reforming. Small mines are still unsafe and account for 75% of the accidents. In the potentially explosive atmosphere of coal mines, open flames—carbide cap lamps—were legal until 1970.[3]

Table 4 shows the hazards involved in producing a given amount of electricity by any one technology. If we produced all of our electricity from coal, the hazard in terms of *deaths alone* would be 20 times that of producing all of our electricity from nuclear power. For the same amount of electricity production, the extraction of coal is 20 times more dangerous than the extraction and purification of uranium. Nuclear power technology is no more dangerous than hydroelectric power. Gas-fired power appears to be the safest of all, but Table 4 does not take into account the greenhouse effect. Gas, like all fossil fuels, produces CO_2 when burned, and would thus contribute to the greenhouse effect, whereas nuclear power would not.

A major problem with fossil fuels is that the Earth is running out of them. With current world proven oil reserves at about 700 billion barrels, oil used at the present world rate of 21 billion barrels a year will last only about another 30 years,[4] or until the year 2020. There is only enough gas, a relatively clean energy source, to last about 50 years. But there is enough coal for 300 years. If we do increase our use of coal, we must learn how to use it cleanly. This is technically possible, but the process is just beginning, and it will be expensive.

Table 4. Estimated Annual Deaths in the U.S. if All Our Electricity Were Generated by Any One Technology[a]

Technology	Deaths	Totals
Fossil fuels[b]		
Air pollution	3000	
Coal		
Black lung disease	1000	
Mining accidents	100	3100
Oil		
Extraction and refining accidents	110	1100
Gas		
Explosions and fires	20	50
Poisoning	30	
Hydroelectric[c]		
Dam failures		
Direct and indirect deaths	170	
Drownings	10	
Other	20	200
Nuclear[d]		
Uranium mining and processing		
Accidents	15	
Cancers	35	
Reactor accidents		
Direct deaths	1	
Cancers	100	150

[a]Deaths per billion megawatt-hours. Numbers are rounded off. Many data are from Richard Wilson, 1975, "Examples in risk-benefit analysis," *Chemtech* **5** (October), 604–607. It must be noted that the numbers shown represent *yearly averages* over a long period of time. Thus, hydroelectric and nuclear accidents are rare but may produce many deaths when they occur.

[b]In 1978–1987, there were an average of 107 deaths per year in coal mining accidents [Mine Safety and Health Administration (MSHA), U.S. Department of Labor]. Only half the coal is used to produce electricity; on the other hand, coal produces only half our electricity—so these considerations balance out.

    I assume that 2000 of the 3000 air pollution deaths would be due to the use of coal, and 1000 due to the use of oil. Estimated from data in L. B. Lave and L. C. Freeburg, 1973, "Health effects of electricity generation from coal, oil, and nuclear fuel," *Nuclear Safety* **14**, 409–428. These data also indicate that the total occupational hazard of oil procurement is about 1/10 that of coal.

[c]Major dam accidents in the U.S. are rare, but minor ones occur frequently. Disastrous dam accidents occur in other countries, such as Italy. The occurrence of such accidents, like Chernobyl, raises the probability of an accident in the U.S.

[d]In 1973–1979, there were an average of 8 deaths per year in the uranium industry ("Uranium Fatals, 1973–79", MSHA, *loc. cit.*). If we were to produce a billion megawatt-hours of electricity by uranium fission, this number would roughly double.

    The accident estimate is based on Chernobyl. We have had no commercial nuclear accidents in the U.S. that have caused death.

Fissionable Atoms

Uranium fission represents an extremely concentrated source of energy. And it is very clean. There are, as we have mentioned, problems with nuclear proliferation and waste disposal, but these are not insurmountable. As for reactor accidents, they are rare, and the next generation of reactors will be much safer to operate than today's models.

There is only enough uranium-235 in the U.S. to last about 100 years, assuming that all of our concentrated power production will be nuclear. However, with breeder reactors, we could produce plutonium-239 fuel from uranium-238, the common isotope of uranium. With breeding, our uranium supply would last nearly 100 times longer, or 10,000 years. Alvin Weinberg, former director of the Oak Ridge National Laboratory, has explained:

> Because the breeder requires little, if any, mining of uranium, its environmental impact is much smaller, at least at the front end of the fuel cycle, than is the impact of the light water reactor. The roughly 300,000 tons of depleted uranium stored outside the diffusion plants [enrichment plants], if used in breeders, could fuel our entire electric system for centuries![5]

Breeder reactors have the further advantage that most modern designs have a lifetime of about 100 years, which would greatly reduce the radioactive waste disposal problem associated with the dismantling and isolation of obsolete reactors.

Aside from the proliferation problem discussed earlier, the only disadvantage of breeder reactors is that they cost more to build than conventional reactors. Whether or not they will become economically feasible in the near future is debatable,[5] but the French Super-Phénix breeder reactor even now produces electricity more cheaply than photovoltaic solar collectors, and it produces that electricity night and day, rain or shine. And of course, while doing this, it is constantly making more reactor fuel than it uses up!

The problem of the disposal of radioactive wastes is often

cited as an argument against nuclear power technology. It is instructive to compare the hazard of nuclear waste with that of waste from other technologies. The radwaste from current U.S. nuclear technology, after holding for 10 years, represents about the same number of lethal doses as that of the arsenic—from manufacturing and use as a pesticide—that we have put on and into the ground. *But the arsenic does not decay,* whereas the radwaste does. After 100 years, the radwaste is less than 10% as lethal as the arsenic. However, because nuclear wastes are closely guarded, whereas arsenic is "routinely scattered around on the ground in regions where food is grown,"[6] the radwaste hazard is in fact far less than indicated above. Arsenic is a stable chemical element—an atom—and there is no way it can decay. The same is true of the heavy metals barium, cadmium, lead, and mercury, some of which are even more poisonous than arsenic in the amounts that manufacturers have put into the ground.

Radioactive atoms do, however, have characteristics that make them more dangerous than other atoms. The radioactive atoms in radwaste are *always mutagenic*—that is, capable of producing inherited changes in organisms—whereas the other atoms are generally much less mutagenic. Also, high-level radwastes are usually *highly concentrated* and may be more dangerous locally. However, once confinement or isolation has been achieved, radwaste is thousands of times less dangerous than these other materials, such as arsenic, that we rarely worry about. We don't worry much about these other materials primarily because they are *familiar* to us.

We should recognize that the radwaste disposal problem is *not* an acute one: we have lots of time in which to solve it. Nuclear power is not growing in this country at anywhere near the rate earlier anticipated, so waste has not accumulated as fast as we thought it would. It is also important to remember that there isn't very much nuclear waste. The total volume of commercial nuclear waste produced in the U.S. during the industry's *entire history* would fit into a small warehouse,[7] whereas the *yearly* volume of chemical waste regulated by the Environmental Protection Agency in 1981 was 264 million metric tons, which would fill the New Orleans Superdome almost 1500 times over.[8]

Finally, there is no harm in letting nuclear waste "sit around" for another decade or so: the longer it sits, the less radioactive it becomes. Furthermore, it is sitting where it poses no danger to the public—unlike most chemical wastes.

Fusible Atoms

Nuclear fusion—the merging of two nuclei of hydrogen to make a nucleus of helium—provides an extremely concentrated form of energy, as demonstrated in the hydrogen bomb. Fusion power is much cleaner than fission power. There are *no radioactive "fusion products,"* like the "fission products" produced by the splitting of uranium, although there would be some releases of tritium. About the only important radwaste fusion does produce—as does fission—is neutron-induced radioactivity in the steel structures around it. Therefore, there is *no severe radwaste disposal problem.* Fusion does not produce transuranic atoms, like plutonium, that can be used in fission bombs, so *proliferation is not a problem.* And, because the "fusion product"— helium—is not explosive, there is *no point in sabotage.* Clearly, nuclear fusion would be a great source of power. But no one has yet been able to harness it.

Fusion is most commonly produced between the heavy hydrogen isotope deuterium ($^2_1$H) and the *very* heavy hydrogen isotope tritium ($^3_1$H). Deuterium makes up 1 out of every 6000 hydrogen atoms in natural waters, including the oceans. As two thirds of the atoms in water (H_2O) are hydrogen, it is evident that there is enough deuterium to last for billions of years. However, not much tritium exists naturally, as its half-life is only 12 years. But it can be produced by bombarding the element lithium (Li) with neutrons from nuclear reactors.[9] The lithium could also actually be put into the fusion reactor, where fusion neutrons would convert it into tritium. The problem is that lithium supplies are limited: high-grade ores may last only a few hundred years.

Much excitement was generated by press reports in the spring of 1989 when electrochemists Stanley Pons and Martin Fleishmann announced that they had achieved "cold fusion," or fusion at room temperature, using an electrochemical cell—something like a car battery. Normally, fusion requires temperatures of 100 million degrees. Most other scientists have not been able to repeat these results, and it now appears that cold fusion was not achieved, although some other energy-producing process may be at work—the jury is still out on this matter.[10]

Of perhaps greater interest is another *confirmed* form of cold fusion, in which negatively charged nuclear particles called *muons*, produced in an accelerator, form tight bonds between the nuclei of two hydrogen isotopes—either deuterium or tritium. The hydrogen nuclei then fuse, releasing the muons for further reaction.[11] The process occurs at room temperature but is optimal at about 900 degrees Celsius, instead of the roughly 100 million degrees required by thermonuclear fusion. A major stumbling block in the development of this form of fusion is that the lifetime of these muons is only about two millionths of a second. However, it has been found that, under proper conditions, a muon may catalyze hundreds of times more reactions before it decays than had previously seemed possible, and it is now at least conceivable that this form of cold fusion may someday become commercially feasible.

We shall probably solve the problem of fusion power by the end of this century—it might then be commercially available as early as the year 2020. Such a development would gradually displace fission power. However, hoping that fusion power will come along before oil runs out would hardly constitute a sound energy policy.

The Power Solution

> . . . . *the aversion people rightly feel for military applications must not spill over to the peaceful use of nuclear energy. Mankind cannot do without nuclear power.*
>
> ANDREI SAKHAROV
> *Memoirs* (Alfred A. Knopf, 1990)

Fusion power would thus seem to be the clear choice, and it is therefore of great importance that we *continue to develop this technology*. One should recognize, however, that the process will probably be expensive, and that it may be another 50 years before it is commercially available. Fusion therefore appears to be only a long-term panacea.

We should also continue the research and development of solar power as a *dispersed* source of energy, particularly for heating homes and schools, and for small, isolated industries.

In the meantime, to what extent should we replace fossil fuels with fission fuels? That is the big question. We still haven't solved the largely political problems of nuclear waste disposal. But that is about the only big problem. Our present reactors are acceptably safe, and we know how to build much safer ones. If we are to become independent of Middle East oil, we should proceed with the further development of nuclear power. We can benefit from the experience of France, which will produce 90% of its electricity through fission power by the year 2000.

As a result of (1) the temporary oil glut, which makes nuclear power less competitive; (2) the inept manner in which some of our nuclear power plants have been constructed, with great delays and cost overruns; and (3) public misapprehensions about the safety of nuclear power, our nuclear power industry has plateaued, and no new power plants are currently being built. We now make nearly 20% of our electricity with nuclear power. Even this relatively low level of nuclear commitment has already mitigated our dependence upon foreign oil—which nevertheless is growing apace, being now about 50%.

Although we use oil for many purposes, such as making gasoline and plastics, nuclear energy is useful only for producing electricity. However, our use of electricity is growing rapidly. Total energy production in the U.S. did not change significantly between 1973 and 1986, but electricity use increased 40%.[7] Furthermore, phasing out our use of fossil fuels could mean moving to electric cars, at least for city use, and the batteries for such vehicles could be charged by the electricity made by nuclear power.

Nuclear power is sometimes criticized because of the "down time," when the plants are shut down for various reasons. Nuclear plants were down—i.e., not producing power—36% of the time in 1977.[12] However, it is often not recognized that fossil-fuel plants have down times of about 22%.[13] In both cases, a major cause is leaks in boiler tubing, a problem not unique to nuclear power. In most cases in a nuclear plant, this tubing carries water or steam that is not radioactive—and when it is radioactive, the activity is very low.

In assessing what we are going to do, we must remember that we are always weighing *risk against benefit*. We expect any concentrated power system to have some hazards. What we have to consider is the *nature* and the *degree* of the hazards. Do we prefer the slight uncertainty associated with burying high-level nuclear wastes deep in the ground to fossil-fuel-produced respiratory disease, the greenhouse effect, and acid rain? The air pollution from burning coal in the USSR produces *in one year* roughly the same number of deaths as will be produced in 50 years by the fallout from Chernobyl.

Lawrence Elliott, an editor at *Reader's Digest*, has provided an eloquent summary:

> The truth is that nuclear energy is a low-risk, high-dread industry. Until Chernobyl, no one had ever been killed by radiation from a civilian nuclear power station. And if we look closely at Chernobyl, the very defects that led to disaster—a Rube Goldberg power plant and an inept oper-

ating crew—ought to give us heart. Western plants are in-
finitely safer.

If the postwar world had turned its back on nuclear
energy, we would have been spared Chernobyl. But the
price would have been decades of still more coal- and oil-
burning, still more acid rain and sick forests, an even worse
"greenhouse effect" over the earth, and thousands more
prematurely dead of respiratory illnesses.[14]

*The most important thing we must do at the present time is to establish
a sound, long-term energy policy.* Richard Helms, former CIA chief
and onetime ambassador to Iran, said:

It is incumbent upon the country—the President and the
Congress—to wrestle with this difficult problem of an energy policy. Otherwise, the United States, in fact the free
world, will become increasingly hostage to unpredictable
events in the Persian Gulf area.[15]

The oil glut was already ending in 1989, before the Iraq crisis,
and oil prices had begun to rise. These prices are now rising
rapidly, and they will stay high regardless of events in the Middle East.[16] Nuclear power is rapidly becoming more economically competitive. We should proceed to build more plants,
using *safer reactors*, and we should promote the design of the
most advanced reactors possible. We should build *power parks*,
containing several reactors, as well as storage ponds and perhaps a reprocessing facility, all at the same site. This must be
done in regions of low population density, with care taken to
site any new plants well away from metropolitan areas, and
downwind of those areas, preferably on a seashore where the
prevailing winds are offshore. Using larger numbers of *smaller*
reactors will lower the probability of accidents, as well as the
environmental impact of any one accident. By 2010, we should
have a permanent high-level waste repository. We can start now
to benefit from the experience of the French, who already have a
nearly total nuclear economy. We should standardize reactor

design, as the French have done; such standardization greatly lowers construction costs and construction time. We need to get over our "fear of flying" and to proceed with a rational nuclear program, including the use of breeder reactors. Even if we choose such a course right now, however, its effect will not be felt until the next century.

We cannot suddenly stop using fossil fuels. But we should be veering away from coal and oil, and using more natural gas for the production of electricity. Gas is better for home heating, where it is much more efficient than electricity. Although burning natural gas produces CO_2 and thus contributes to the greenhouse effect, the gas is otherwise clean, and is far preferable environmentally to coal or oil. We shall undoubtedly continue to produce much of our concentrated power with coal, in large part because we have such great reserves of it. Objectionable as coal is in terms of its environmental pollution, systems are being developed that will make it much less polluting. Such developments, of course, will make coal-produced electricity more expensive and, in turn, will make nuclear power more attractive.

IV ⚛ The Peril

12 ⚛ Nuclear Weapons and Arsenals

We turn now from peace to war. Controlling the fission of uranium atoms enabled humankind to produce a broad arsenal of radioactive isotopes, which are used in myriad ways to advance the science and technology of medicine, agriculture, and manufacture. Such controlled fission has also opened the door to clean, safe, and nearly limitless energy to run the machines of modern society. But the release of nuclear energy has also been a Pandora's box. Evil spirits have emerged along with the good, and we now know how to destroy civilization within the span of a single day. Such incredible power is without precedent in human history. *Its consequences must be squarely faced.* We begin by looking at the nuclear weapons themselves.

Some atoms of uranium and plutonium undergo *spontaneous fission*, meaning that, with a very low probability, their nuclei will spontaneously split apart, releasing several neutrons and a tremendous amount of energy. Occasionally, one of the released neutrons will strike another nucleus and produce an *induced fission*. If a sufficiently large mass of uranium is treated (enriched) so that the concentration of the fissionable isotope uranium-235 increases, the neutron density from spontaneous fission will rise until induced fission becomes the predominant process. If the enrichment increases, one will soon reach a *critical mass* of U-235, and a chain reaction will begin, producing tremendous heat. If the chain reaction is not immediately stopped, the mass will explode.

In a nuclear reactor, the density of neutrons is very carefully controlled, so that the reactor just barely exceeds criticality. A lot of heat is given off from the controlled fission that ensues, and that heat is used to produce steam that drives turbines, which in turn drive electric generators. In a reactor that uses uranium, the concentration of the critical isotope, uranium-235, is typically about 3%. In a uranium fission bomb, on the other hand, the U-235 concentration is at least 90%, and—by bringing subcritical masses together very rapidly—a highly supercritical mass is created in a fraction of a second. An extensive chain reaction then occurs, producing a tremendous explosion. The entire chain reaction takes less than a millionth of a second.

Bomb Types

Fission Bombs

A nuclear fission bomb, or A-bomb, can be made only of a material with a high atomic weight, whose atoms will undergo induced fission by neutrons to produce a chain reaction. The most practical isotopes for nuclear bombs are uranium-235 and plutonium-239—the materials used in the bombs dropped on Hiroshima and Nagasaki, respectively.

Because a nuclear bomb explodes so quickly, the change from subcritical to supercritical mass has to occur very rapidly; otherwise, the bomb would begin to explode before getting very supercritical, and it would fizzle. The rapid transition to supercriticality may be achieved in a variety of ways. One, used in the Hiroshima bomb (Figure 29), is to have a TNT explosion drive one subcritical hemisphere of U-235 against another one, along what is effectively a gun barrel. When the two hemispheres come together, they form a supercritical mass, which then explodes. The more usual technique, which was used in the

FISSION BOMB

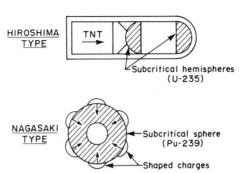

Figure 29. Simplified schematic diagram of the construction of fission bombs (A-bombs).

Nagasaki bomb, is to produce an *implosion*, created by conventional explosives on the surface of a subcritical sphere (of low density or with a hollow center), that compresses the sphere into a supercritical mass. Microsecond timing of the explosions is required. One of the problems that terrorists would have in making a successful nuclear bomb would be creating a supercritical mass with sufficient speed and precision to make a big bang instead of a little poof. This requirement also provides one reason why a nuclear reactor—which is designed to be just barely critical—cannot explode like a nuclear bomb.

Uranium and plutonium bombs are called nuclear fission bombs because they involve the fission of nuclei. They were erroneously called "atomic" bombs when they were first used in World War II. Nevertheless, the term "A-bomb," meaning "atomic bomb," has persisted as a term referring to fission bombs.

As the critical mass is achieved in a bomb, the neutrons released by spontaneous fission can be greatly augmented by an *initiator*, which is a device that produces a sudden spray of neutrons when the bomb is triggered. This spurt of neutrons gets the chain reaction going quickly, resulting in more fission—and thus a higher yield—before the bomb blows itself apart.

Largely because the fission of an atom of plutonium-239 produces up to *three* neutrons, instead of the two neutrons produced by U-235 fission, plutonium is the favored material for fission bombs. This plutonium is made in *plutonium production reactors* operated by the military, and it is used in bombs in a nearly pure form. It is also produced, quite incidentally, in the normal operation of a nuclear power reactor, which is the reason why the export of power-reactor technology, and especially breeder-reactor technology, is carefully monitored by international agencies in an effort to prevent smaller nations from acquiring stocks of bomb-grade plutonium.

The Nagasaki bomb contained about 8 kilograms of plutonium, which is about the size of a grapefruit. The temperature generated when such a bomb explodes is over 100 million degrees Celsius, and the accompanying pressure is millions of times greater than normal atmospheric pressure. These are comparable to the conditions that exist at the center of the Sun.

Fusion Bombs

A totally different way of making a nuclear bomb involves the *fusion* of two atoms of *very low* atomic weight. Two hydrogen atoms, for example, can be made to fuse so that they form a helium atom, as shown in Figure 15. This process, called "nuclear fusion," releases tremendous amounts of energy. Because hydrogen is used, nuclear fusion bombs are called "hydrogen bombs," or H-bombs. Such fusion requires extremely high temperatures, and weapons using it are therefore also called **thermonuclear weapons**. The only isotopes of hydrogen that are practical for a fusion bomb are deuterium (^{2}H) and tritium (^{3}H).

Fusion bombs use a fission bomb for a trigger! This is the only way to produce the tremendous temperature and pressure required to ignite the fusion process. In principle, a hydrogen bomb (Figure 30) involves an implosion-type fission bomb

FUSION BOMB

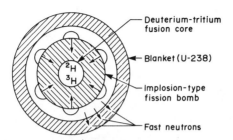

Figure 30. Simplified schematic diagram of the construction of a fusion bomb (H-bomb). A conventional implosion-type fission bomb ignites the fusion bomb core; fast neutrons from both explosions then produce further fission in the U-238 blanket. Current fusion bombs are generally much more complex than the one shown here.

placed around a core of deuterium and tritium. Upon implosion, both the fission and the fusion elements are compressed. When the fission bomb explodes, it creates sufficient additional heat and pressure to fuse the deuterium and tritium, causing a tremendous release of fast neutrons. The whole thing is surrounded by a spherical shell of U-238, called a **blanket**, in which further fission is induced by the fast neutrons.[1] A thermonuclear weapon is therefore a fission-fusion-fission bomb. In typical bombs, roughly half the output is due to fission reactions and half to fusion reactions.

The size of fission bombs is limited by the critical mass to a maximum yield of about 50 **kilotons**. By this, we mean a bomb with the *explosive force* (not weight) of 50,000 tons of TNT. For perspective, we may note that the largest aerial bombs used in World War II carried 10 tons of TNT. Figure 31 shows that the explosion of such a fission bomb is an awesome phenomenon. But there is no theoretical limit to the size of a fusion, or hydrogen, bomb; the Soviet Union has exploded the largest one: 58 **megatons**. This is a bomb with the explosive force of 58 *million* tons of TNT! A 10-megaton bomb will destroy any city on Earth. Ten megatons of TNT would fill a freight train stretching from

Figure 31. Shallow underwater explosion of a 30-kiloton fission bomb at Bikini Atoll in the South Pacific in 1946. The explosion dwarfs full-sized naval vessels. (Courtesy of U.S. Department of Energy.)

Los Angeles to New York. Such tremendous bombs are impractical and are no longer made, because several smaller bombs, mostly no larger than 1 megaton, are more effective. But even a "mere" 1-megaton bomb produces an explosion *fireball* that is over a mile in diameter.

Neutron Bombs

A **neutron bomb** is basically a hydrogen bomb without the outer U-238 blanket. The tremendous numbers of neutrons released by the fusion process, instead of being absorbed by the blanket, are therefore released from the bomb and sprayed onto the ground.

The purpose of a neutron bomb is to produce acute radiation sickness in soldiers, and less destruction of property. Fast neutrons can penetrate tank armor. It may appear to be a less awful weapon than a hydrogen bomb, because the explosion is smaller, producing less destruction from the blast and the heat. However, on most battlefields, the neutrons will kill large numbers of civilians. It has been estimated that such a bomb used in Germany would kill 200 civilians for every tank crew immobilized.

Nuclear Artillery

Although countries beginning a nuclear weapons program might build bombs similar to those used on Hiroshima and Nagasaki, these bombs are crude when compared with modern American and Soviet warheads. The Nagasaki bomb weighed 5 tons and had an explosive force of 20 kilotons. A modern American strategic warhead weighs about 200 pounds and has an explosive force of 350 kilotons. The explosive part of the Nagasaki bomb weighed 8 kilograms and was about the size of a grapefruit, but we can now make a critical mass with 2 kilograms (4.4 pounds) of plutonium—a volume smaller than a tennis ball. A nuclear fission warhead with the explosive power of the Nagasaki bomb can now be placed inside a 6-inch, or 155-millimeter, artillery shell, and a neutron bomb can be put inside an 8-inch artillery shell.

Strategic Weapons

ICBM

The intercontinental ballistic missile (**ICBM**) is the most important weapon in any nuclear arsenal. This weapon is typically a multistage solid-fuel **rocket**, which carries a nuclear bomb, called a **warhead**. These weapons are termed **ballistic missiles** because their rockets merely get them started, much as the gunpowder in a cannon gets the shell going. Once above the atmosphere, the rockets stop firing and fall away, and the missile coasts all the way to its target along a "ballistic" trajectory, a trajectory affected only by gravity and the velocity of the missile—just as a shell from a cannon follows a ballistic trajectory once it has left the gun. Such missiles, launched from underground **silos** located mostly in the Midwest of the United States, have ranges up to 7000 miles and can be flown to any part of the Soviet Union. The travel time is about 30 minutes. The heat and light given off at launch can be easily detected by satellites, so target populations can expect about 25 minutes' warning. The accuracy is astounding. The best American missiles can strike within a hundred yards—the length of a football field—of their targets. Such accuracy is sufficient for attacks on cities or airfields, but marginal for attacks on missile silos or command posts that have been "hardened" against attack— these sites would be destroyed only by a direct hit. Radar-equipped warheads are now being developed that will have much better accuracy.

Soviet missiles are less accurate than American missiles, and they are generally less reliable and efficient, requiring a larger rocket to carry equivalent warheads.

MIRV

A *m*ultiple-headed *i*ndependently targetable *r*eentry *v*ehicle (**MIRV**) is simply an ICBM that has as many as 10 warheads on a single missile, *each warhead programmed for a different target*. After the missile has been propelled into space, and the booster rockets have fallen away, the front part of the rocket, called a *bus*, releases its independently targeted warheads. A single MIRVed ICBM could either concentrate its warheads on a metropolitan area, with devastating effect, or spread its warheads out and destroy several separate smaller cities.

The standard American ICBM, called the *improved Minuteman III*, has three warheads, each with a 335-kiloton explosive force, totaling about 1 megaton for each missile (see Table 5). The

Soviets have roughly comparable missiles, but generally with more powerful warheads. The biggest American ICBM is the *MX missile*, which carries 10 warheads of 300 kilotons each, for a 3-megaton total. The Soviet *SS-18* and *SS-24* are somewhat larger weapons, with 10 warheads of 550 kilotons each, for a total of 5.5 megatons.

We noted above that the largest nuclear bombs tested have rarely been made into operational weapons because they are so wasteful. Few warheads now have a power exceeding 1.2 megatons; most strategic warheads are currently less than 600 kilotons. It is quite sufficient to sprinkle a metropolis with ten 300-kiloton bombs, each 20 times as powerful as the Hiroshima bomb!

There is some tendency today to move away from MIRVing. One of the major concerns of military strategists is that the enemy will attempt to knock out our missile silos first, in an effort to prevent retaliation. It is therefore desirable to have a *highly mobile ICBM*, but our present weapons are too bulky and heavy for easy mobility. The current American solution is the *Midgetman*, a single-warhead missile less than 4 feet in diameter and 38 feet long. The Midgetman could be carried on the flatbed of a low-profile, blast-resistant truck, and would be capable of quick launch by remote control. Even if a Soviet satellite should detect the exact location of a Midgetman before giving a command for the firing of an ICBM from the USSR, the truck would have 25 minutes to move to another location, which could be done at 50 miles per hour on good roads.

SLBM

A *submarine-launched ballistic missile* (**SLBM**) is in principle the same as an ICBM. Because of size constraints, SLBMs tend to be smaller than land-based ICBMs and to have a shorter range. They are designed so that they can be launched underwater by a blast of steam, which projects them right out of the water, whereupon their rocket fires and they climb into the air like any other missile. Most of them are now MIRVed. The Soviet *SS-N-8* is a single-warhead 1.5 megaton SLBM, but the latest models are the American *Trident* and the Soviet *SS-N-23*, both MIRVed and both having 100-kiloton warheads.

Clearly, the range of an SLBM generally does not have to be as great as that of an ICBM, as the submarine can be stationed near enemy shores. This shorter range also shortens the warning time. Enemy subs off our east coast, firing missiles on New York and Washington, would give us no more than 5 minutes of warning, even if we detected them at the time of launch.

A modern Trident sub has 24 missiles, each with 8 warheads, for a total of 192 warheads. The accuracy of these missiles is comparable to the moderate accuracy of a Minuteman III. Although they would be relatively ineffective against hardened missile silos and command posts, two such subs, in either the North Atlantic or the Indian Ocean, could destroy every major city and military base in the Soviet Union. Our current Trident subs carry the Trident I (C-4) missile. These are now being displaced by the far more accurate Trident II (D-5), which could knock out missile silos and thus have a first-strike capability. Such weapons dangerously upset the nuclear "balance" of the U.S. and Soviet forces.

Aerial Bombs

Nuclear weapons were originally carried by bomber aircraft and were deployed like any other aerial bomb. Although bombers in the U.S., the USSR, Britain, and France continue to carry strategic nuclear weapons, they now play a lesser role, because they are relatively slow compared to ballistic missiles, and they can be shot down.

Tactical Weapons

The ICBMs and SLBMs are clearly **strategic nuclear weapons**, that is, weapons used in the overall strategy of defeating an enemy, and they are usually aimed at the enemy homeland.

Tactical nuclear weapons are used on the battlefield. Included among tactical weapons are some of intermediate range that can play either a tactical or a strategic role. The military call these *long-range theater nuclear weapons*, but they may also be called **intermediate-range weapons** (range 600 to 3000 miles), as distinct from *shorter-range weapons* (300 to 600 miles), which are exclusively tactical. Most of these weapons are mobile.

The American Pershing II is an intermediate-range missile that was deployed in Europe until recently. It is a two-stage solid-fuel rocket with a range of about 1500 miles, which would have been sufficient to hit any part of European Russia when fired from Germany. The flight time (less than 10 minutes), and hence the warning time, is much shorter than for an ICBM. The deployment of this weapon in Europe was therefore viewed with alarm by the Soviets. It has now been removed under the Intermediate-range Nuclear Forces (INF) treaty, signed by President Reagan and General Secretary Gorbachev in 1987.

One of the most important tactical weapons is the **cruise missile**, which is essentially a flying bomb, not unlike the "buzz bombs," or V-1's, of World War II. The name is a bit of a misnomer: a cruise missile is not a missile in the normal sense of an arrow or an ICBM. It is instead a "pilotless aircraft." A cruise missile can be launched from the ground, or from a ship or a bomber. It has stubby wings and a jet engine, and flies at about 500 miles per hour, but within a few hundred feet of the ground, thus being very difficult to detect or track, either by radar or by satellite. It has a radar sensor in its nose, which observes the terrain and compares it with a strip map on which its course has been plotted. It is very accurate. The American cruise missiles recently deployed in Europe had ranges up to 1500 miles; used at this range, they are a strategic weapon. They too were eliminated by the INF treaty.

Submarine-launched missiles can be used as an intermediate-range weapon, as can bomb-carrying aircraft. Ballistic missiles, cruise missiles, and aircraft can all be used at shorter ranges, if desired.

Short-range weapons, which travel less than 300 miles, are typically rockets with only a single solid-fuel stage. They can be used as *ground-to-ground, ground-to-air, air-to-air,* or *air-to-ground* missiles. These weapons are not ballistic missiles, which coast through much of their trajectory, but true rockets, which fire throughout the trajectory.

Cruise missiles and short-range rockets with nuclear warheads are now standard weapons on warships. The American Tomahawk missile is a *ship-to-ship* or *ship-to-shore* nuclear-capable rocket with a range of about 300 miles. The French Exocet missile, which was used by Argentina in 1982 to sink the large British destroyer *HMS Sheffield,* was not a nuclear weapon; it nevertheless demonstrated how modern rockets may be rendering the heavy guns of warships obsolete.

A wide variety of *nuclear artillery* is now available. Nuclear warheads with yields from 1 to 20 kilotons, including neutron bombs, can be fitted into artillery shells.

Ballistic missiles, incidentally, need not have a nuclear warhead; they may carry ordinary explosives. Missiles with TNT warheads or chemical-warfare warheads are now designed to travel up to several hundred miles. Iraq fired TNT ballistic missiles at Teheran during the Iraq-Iran war, but they did relatively little damage.

Nuclear Arsenals

The following is only the briefest description of the nuclear arsenals of the world. The size and composition of the arsenals is constantly changing, but the general capabilities of the nuclear powers are remarkably well known, in spite of military security. It is not necessary, however, to have detailed knowledge of these arsenals in order to arrive at rational conclusions about super-power nuclear capabilities.

Table 5 shows that the numbers and characteristics of the strategic nuclear missiles on both sides are roughly comparable. The Soviets have somewhat greater numbers of missiles, with

Table 5. Strategic Nuclear Missiles of the United States and
the Soviet Union at the Beginning of 1990[a]

Missile	Characteristics			No. of missiles
	No. of warheads ×	kton/warhead =	kton/missile	
ICBM				
U.S.				
Minuteman II	1	1200	1200	450
Minuteman III	{ 3	170	500 }	500
	3	335	1000	
MX	10	300	3000	50
				1000
USSR				
SS-11	{ 1	1100	1100	150
	3	350	1000	210
SS-17	4	750	3000	100
SS-18	10	550	5500	310
SS-19	6	550	3300	300
SS-24	10	550	5500	60
SS-25	1	550	550	170
				1300
SLBM[b]				
U.S.				
Poseidon C-3	10	40	400	210
Trident I C-4	8	100	800	380
				590
USSR				
SS-N-6	2	1000	2000	190
SS-N-8	1	1500	1500	290
SS-N-18	7	500	3500	220
SS-N-20	10	200	2000	120
SS-N-23	4	100	400	100
				920

[a] *Bulletin of the Atomic Scientists* **46** (1990) January/February, p. 49; March, p. 49; September, p. 49. Figures are rounded off.
[b] At the beginning of 1990, the U.S. had 33 nuclear submarines, 13 carrying 16 Poseidon missiles each, plus 20 carrying up to 24 Trident missiles each, for a total of 590 MIRVed missiles. The USSR had 61 nuclear submarines, of which 38 carried from 16 to 20 MIRVed missiles, for a total of 630 MIRVed missiles.

larger warheads. For example, the newest and most threatening missile in the American arsenal is the MX, with 10 warheads of 300 kilotons each—we have 50 of these missiles. The newest Soviet missile is the SS-24, with 10 warheads of 550 kilotons each—they have 60 such missiles. Our most modern submarine, the Trident, carries 24 missiles, each having 8 warheads of 100 kilotons power. The latest Soviet Delta IV subs carry 16 SS-N-23 missiles, each having 4 warheads of 100 kilotons power. Although the total number of Soviet submarines, and submarine missiles, is considerably greater than that of the US, the number of submarine MIRVed missiles of the two nations are roughly equal (footnote, Table 5).

Table 6 shows that the U.S. has the same number of strategic warheads as the USSR, although the total *megatonnage*—millions of tons of explosive force—of Soviet missiles is greater. The reason is that a majority (57%) of the Soviet weapons are ICBMs, and relatively few (11%) are carried by heavy bombers. This breakdown is illustrated in Figure 32. The Soviets have many of

Table 6. Total Numbers and Megatonnage (MT) of Warheads in the World's Nuclear Arsenals[a]

	Strategic[b]	Tactical	Total
U.S.	12,000 (3000 MT)	9000 (1500 MT)	21,000 (4500 MT)
USSR	13,000 (6000 MT)	17,000 (5000 MT)	30,000 (11,000 MT)
			51,000 (15,500 MT)
China	300 (470 MT)		
France	620 (135 MT)		
UK	300 (60 MT)		
	1220 (665 MT)		

[a] Approximate values. Data for the U.S. and the USSR from *Bulletin of the Atomic Scientists* **44** (1988): January/February, p. 56; **46** (1990): January/February, p. 49; July/August, p. 49. Data for China, France, and the United Kingdom, which include tactical weapons, from *Bulletin of the Atomic Scientists* **46**, (1990): November, p. 49; December, p. 57.
[b] Includes bombs and cruise missiles carried on long-range aircraft. Soviet strategic defense weapons (antimissile missiles used in defense of cities) are listed under "Tactical"; the U.S. has deployed no such weapons.

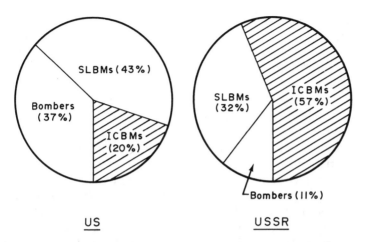

Figure 32. Comparison of the proportions of strategic nuclear warheads in various delivery systems, for the U.S. and the USSR at the beginning of 1990. (Data from *Bulletin of the Atomic Scientists* **46**, 1990: January/February, p. 49; March, p. 49.)

their land-based missiles directed at Europe and China, whom they have considered potential enemies. The U.S. has a much more balanced force, with 20% in ICBMs, and roughly 40% each in SLBMs and heavy bombers. The U.S. has concentrated its force in the so-far-invulnerable nuclear sub—a deployment consistent with our presence in most of the oceans of the world. Only about half of a fleet of nuclear missile subs can be on patrol at sea. The remaining ones in port are vulnerable to attack, although they are not prevented from firing their missiles before being attacked. The Soviet sub force is somewhat bottled up in the Sea of Japan, the Barents Sea, the Baltic Sea, and the Black Sea, all of which can be blockaded by American or NATO forces; thus, submarine deployment is less attractive to the Soviets.

Although the U.S. and the USSR now have roughly equal numbers of strategic warheads (Table 6), in the past the U.S. has always had superiority (Figure 46, Chapter 20). Soviet weapons have generally had greater power (megatonnage), but the American weapons are more accurate. Thus, the many claims over the years by politicians and the military of "windows of vulnerability" and "missile gaps" have been little more than propaganda.

The Soviets have almost twice as many tactical missiles as the U.S. (Table 6), but most of these are deployed along the border of the Soviet Union, while the U.S. has no need for such a tactical deployment on its native land. Many of the tactical missiles of both nations are on ships.

The nuclear arsenal of Great Britain is puny in comparison to those of the two superpowers, yet the force of Britain alone—about 60 megatons, all on aircraft and subs—is 4000 times the force of the Hiroshima bomb, and 10 *times the total tonnage of weapons of all types exploded by all countries in World War II.* The four nuclear subs of Great Britain have twice the power necessary to destroy every important city and military base in the Soviet Union. Finally, the British plan to replace their present subs with four Trident-class subs, and to double their numbers of warheads by 1995.

France has twice as many nuclear warheads as Britain, and has six nuclear subs. France also plans future enlargement. China has about the same number of warheads as Britain, but much greater megatonnage. China has two nuclear subs.

The nuclear forces of the world at the beginning of 1990 comprised 52,000 warheads, with a total of 16,000 megatons of explosive force. This is about 3 tons for every man, woman, and child in the world. If these weapons were ever all to be exploded in a short period of time, it would certainly be the end of civilization and possibly the end of humankind. Any single modern nuclear sub can devastate a large nation.

One might think that the near collapse of the Soviet communist world would have put an end to nuclear armament. But nuclear arsenals on both sides have not been lowered by more than 5%. And the Bush administration still pushes for development of more sophisticated nuclear weapons in general, and of the Strategic Defense Initiative (Star Wars) in particular. One may hope that future START talks will make some truly significant reductions. And much else can be done, along the lines of test bans. We discuss these prospects in later chapters.

Some may still raise the question of whether or not the higher numbers in Tables 5 and 6 for the USSR are significant. President Reagan seemed to answer this for us on 18 December 1982, early in his administration. When asked, "Looking at U.S. defenses overall, would you trade American forces for Soviet forces?" he replied "No."

13 Nuclear War
I. The Terrible Swift Sword

The prospect of apocalypse strips words of their ordinary meanings. We recognize that the detonation of only a single medium-sized thermonuclear weapon over a major metropolitan area in the United States (or any other nation) would produce death, injury, destruction, and devastation on a scale that is without precedent in the history of mankind.

H. Jack Geiger, M.D.
Last Aid (W. H. Freeman, 1982)

The world has never had a nuclear war. Therefore, we do not know exactly what would happen. But if the major powers engaged in a nuclear war, it probably would be the last for at least a thousand years, because human civilization would be destroyed. Preventing nuclear war is the most important issue facing humankind today. On the road toward prevention, we must first do our best to estimate what would happen in a nuclear war.

The bombings of Hiroshima and Nagasaki were frightful events. One may debate the morality of these acts, but it can be argued that they were justified, in terms of saving lives, both

215

Japanese and American, because they immediately ended the war. Humankind may have gained some insight from this tragedy. Hiroshima and Nagasaki could be considered gruesome experiments that gave us a "micropicture" of nuclear war. Without this frightening example, we might not have been able to stave off nuclear war as long as we have.

The Hiroshima bomb had a power of about 15 kilotons (15,000 tons) of TNT, and the Nagasaki bomb had a power of about 21 kilotons. A single modern American nuclear submarine carries almost 200 warheads with a total explosive force of 20 megatons (20 *million* tons) of TNT. This is 1000 times the force of a Hiroshima or Nagasaki bomb. Any one of these submarines could devastate most of the large cities and military bases of either the U.S. or the USSR. The United States has the equivalent[1] of about 25 such modern subs and the Soviet Union a similar number.

The United States and the Soviet Union each have a total of over 10,000 strategic warheads. Assuming that 200 would devastate and paralyze either nation, and that 1000 would utterly destroy that nation, our total strategic arsenals represent **overkill** by at least a factor of 10.

This estimate does not include our tactical nuclear weapons, many of which could also be used as strategic weapons against the enemy homeland.

War Scenarios

The following scenarios involve the U.S., the Soviet Union, and the Warsaw Pact nations because the nuclear arsenals were designed with such conflicts in mind. As long as the arsenals continue to exist, such war is possible, although perhaps with different protagonists.

Most analysts assume that, in an all-out nuclear war, the superpowers would manage to deliver only about half of their strategic nuclear weapons. Yet, in such a war, at least 10,000 strategic warheads would be released, representing some 3000 megatons (MT). If Europe were involved, another 3000 MT of tactical weaponry would be delivered. Thus, an all-out nuclear war could involve the release of some 6000 MT of explosive force.

European War

We might see a *limited nuclear war*, in which relatively few bombs were exploded. One can imagine that an attack on Europe by the Soviet Union would provoke a nuclear attack by the U.S. on the battlefield only, perhaps followed by Soviet nuclear attacks on a few cities in France and England. At this point, the two superpowers might come to their senses and, frightened by the prospect of escalation, call off all further nuclear exchange. Most analysts consider such a scenario unlikely, partly because a NATO nuclear response may at present be the only way that a Soviet invasion could be stopped, and partly because the fear and hatred engendered by aggressive military action does not usually lead to such a cool reaction. We must remember that Europe is not yet denuclearized. Even with the Intermediate-range Nuclear Forces (INF) Treaty, 4000 nuclear warheads remain in NATO forces.

U.S.–Soviet War

Any attack on the homelands of the U.S. or the USSR would probably be preceded by the explosion of several nuclear weapons, launched by subs, high over the landmass of the enemy country. Such high-altitude explosions can produce an *electromagnetic pulse*, a wave of electrical energy that can destroy electronic communications. One could also expect an early nuclear-sub attack on the enemy capitol, a Soviet attack being launched from the Atlantic and an American attack from the North Sea. Such attacks would provide virtually no warning.

A "somewhat limited" kind of nuclear war might occur.

This would be an attack solely on the nuclear forces of the other nation. Thus, the Soviets might engage in a **first strike** against our ballistic missile silos, military command posts, nuclear sub bases, and Strategic Air Command (SAC) airfields, in the hope that enough of our arsenal could be eliminated at the outset that we would be unable to make an effective response. This scenario, called a **counterforce** attack, also does not seem very likely, as our nuclear subs at sea would be immune and could respond with devastating effect.[2] An attack on cities only is called a **countervalue** attack. A large nuclear war would probably involve both counterforce and countervalue attacks.

Let us now consider the physical and human consequences of nuclear war. It must be recognized that the immediate effects are only about half the total consequence. Unlike conventional war, a nuclear war also produces *delayed* effects that may be even more devastating than the immediate ones.

Immediate Effects: Physical

What happens when a nuclear bomb explodes? The events are complex, resulting from energy in three basic forms: blast, heat, and radiation.

Heat is simply the motion of atoms and molecules. In a hot object, be it a frying pan, the air in a warm room, or your own skin on a hot day, the molecules are moving faster. In a cold object, they move slower. Fast-moving molecules exert more pressure on their container, so if the air in a balloon is heated, the balloon expands.

In a nuclear bomb, the tremendously fast-moving particles—fission products, neutrons, and electrons—produced during the explosive chain reaction collide with the atoms in the air in which the bomb explodes, imparting a high velocity to them. In addition, the X rays and many of the gamma rays given off by the nuclear fission are absorbed by air molecules, also causing them to move faster. These two effects together almost instantly heat the surrounding air to tremendously high temperatures, creating the *fireball* associated with all atmospheric nuclear explosions.

Blast

This tremendous production of heat in a small volume produces a very strong pressure wave that we call a **blast wave**. This blast wave moves out at the speed of sound, behaving like an extremely strong wind, and will shatter all nearby objects, including buildings. Fortunately, the blast intensity falls off at a rate somewhere between the square and the cube of the distance from the explosion, so that the effect at twice the distance will be only about one-sixth as strong. Table 7 shows the effects of the

Table 7. Effects of the Blast Wave from a Nuclear Explosion
(1-megaton burst at 1-mile altitude)[a]

Peak overpressure (psi)	Effects	Distance to which effects are felt (miles)
20	Multistory reinforced concrete buildings demolished. *500 mph wind.*	1.8
10	Most factories and commercial buildings collapsed; wood and brick apartments destroyed. *300 mph wind.*	2.7
5	Heavier construction badly damaged. Brick and wood houses destroyed. *160 mph wind.*	4
2	People injured by flying glass and debris. Severe damage to houses. *70 mph wind.*	7
1	Light damage to commercial structures. Moderate damage to houses.	10

[a]From Table 2 in Leo Sartori, 1983, "Effects of nuclear weapons," *Physics Today* (March).

blast wave from a 1-megaton bomb exploded at 1 mile altitude. Virtually every structure will be destroyed within a 2.7-mile radius of *ground zero*, which is the point on the ground directly beneath the bomb. Frame houses will be destroyed out to a radius of about 4 miles, where there is an *overpressure* of 5 pounds per square inch (psi), meaning 5 psi above the normal atmospheric pressure of 15 psi. (Figure 33 shows the effects from a much smaller explosion.) The blast at this distance will hurl a standing human against a wall with several times the force of

Figure 33. In this 1953 test of the effects of a 15-kiloton fission explosion at Yucca Flat, Nevada, a wooden-frame house, two-thirds of a mile from ground-zero, experiences 5 psi (pounds per square inch) overpressure. A blinding flash lights up the front of the house at the moment of explosion (*above*), and the paint (not the wood) of the house catches fire almost immediately (*top right*). About 2 seconds later (*bottom right*), the blast wave arrives and demolishes the house. Note the roof "peeling up." (Courtesy of U.S. Department of Energy.)

gravity. Glass shards, stones, metallic objects, and anything else shattered by the blast, will fly about at speeds of 100 miles per hour, and many people will be mortally injured by these flying objects. At 7 miles, there will still be hurricane-force winds. Clearly, the blast effects alone of such an explosion will be devastating.

Heat

About half of the heat energy of the fireball goes into producing the blast wave. The other half is simply propagated as a wave of radiant heat, referred to as the **thermal pulse**. The thermal pulse lasts from about 1 second for a 15-kiloton Hiroshima-size bomb to *10 seconds* for a 1-megaton bomb. Ten seconds is a very long time to be exposed to a source 1000 times as bright as the sun. Clothing will catch fire out to about 5 miles from a 1-megaton burst, and third-degree burns will be experienced out to 7 miles. Any creature within 25 miles that looks directly at the fireball—a natural reflex action—even briefly, will be permanently blinded. Within 5 miles of ground zero, virtually all combustible materials will burst into flame, although wood-frame houses may only char, because the blast wind will blow out some fires. In addition to the thermal pulse, which passes in seconds, the very hot air of the fireball will continue to ignite fires for several minutes after the explosion—a minute is a long time when one is exposed to the heat of a blast furnace.

The wide spread of fire within 5 miles of ground zero will probably start a **firestorm**. The great heat from the burning area will cause air to rise rapidly, sucking in more air from outer regions and producing winds of hurricane force that will fan the fires and may keep them going for days. Such firestorms were produced in World War II by the saturation bombing of Hamburg, Dresden, and Tokyo with non-nuclear incendiary and blast bombs. Few people survived such firestorms, least of all those hiding in underground "shelters," who were either incinerated or died from asphyxiation.

In addition, many synergistic effects will occur. The blast

alone can cause fires, by overturning stoves and furnaces, and by causing electrical short circuits. The blast will also break gas mains and fuel tanks, which will then be ignited by the heat. Much of the debris created by the blast will be flammable.

Radiation: Immediate

The effects described so far do not differ in principle—although they differ greatly in magnitude—from those that would be produced by conventional explosives such as TNT. The blast and thermal effects of a 100-kiloton bomb would be much the same as from exploding 100,000 tons of TNT. Indeed, tests to measure the blast effects of small nuclear weapons have sometimes been simulated by exploding large quantities of TNT. Unique to the nuclear weapon, however, is the emission of nuclear radiation. This adds to the usual explosion effects, but it does not add as much as one might think.

Gamma rays and neutrons are the primary types of damaging radiation emitted by nuclear bombs. Electrons and other charged particles are absorbed by air molecules within a few meters of the explosion. Fission products rise with the fireball and become part of the **fallout**. Most of the gamma rays and neutrons are also absorbed near the explosion, and they contribute to the heat effects described above. However, a good deal of the gamma and neutron radiation is sufficiently penetrating to reach the ground and can become biologically significant. But these radiations fall off very rapidly with distance, because of absorption in the air. For example, the gamma-ray dose at 1 mile from a 1-megaton blast is about 25,000 rems. At 2 miles, however, it is down to about 40 rems, which is $1/10$ the lethal dose for humans. The practical consequence is that *anyone exposed to a lethal dose of radiation from a nuclear explosion will probably be killed anyway by the blast or the heat*. For example, only about 4% of the *survivors* at Hiroshima and Nagasaki got doses greater than 100 rems. People beyond the blast-and-thermal region may get high doses of radiation, but much of this would come from fallout during the first hour or so after the blast.

The neutrons from a fission bomb, although typically only about $1/10$ as numerous as the gamma rays, are roughly 20 times as damaging biologically (Appendix B). Therefore, the biological effects of gamma rays and neutrons from fission weapons are very roughly the same. For fusion bombs, the neutron component is higher. It should be borne in mind, however, that the relative number of neutrons depends greatly on the nature of the bomb and the altitude of detonation. At Hiroshima and Nagasaki, almost all of the radiation damage was due to gamma rays. Both neutrons and gamma rays penetrate deeply into human tissue and can be lethal to those near the explosion. However, as we discussed in Chapter 3, the shielding parameters are quite different. Someone inside a steel structure will be well-shielded against gamma rays but not neutrons, while someone inside a concrete building will be well-shielded from the neutrons but not the gamma rays.

The neutrons—but not the gamma rays—from the explosion itself will cause **induced radioactivity** in the atoms of most things they encounter, such as building materials and soil. Therefore, an area exposed to a nuclear blast will become radioactive and may be dangerous to humans for quite some time. This induced radioactivity, however, may produce less biological damage than the radioactivity produced by early fallout, for most survivors will manage, one way or another, to leave the immediate blast area within a day.

Finally, mention should be made of the **electromagnetic pulse** produced by nuclear explosions in outer space (Figure 34). If a nuclear bomb is exploded above the atmosphere, the gamma rays move out in a spherical wave at the speed of light. When this wave strikes the atmosphere, the gamma rays are absorbed by air molecules, which become ionized. The separation of electric charge during ionization is very sudden, and it produces a strong electromagnetic wave similar to a radio wave, but one in which there is a tremendous voltage gradient (change of voltage over a short distance), amounting to tens of thousands of volts per meter. This electromagnetic pulse can severely damage elec-

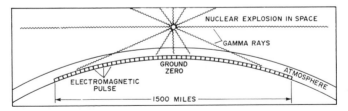

Figure 34. Electromagnetic pulse generated by a high-yield nuclear explosion 100 miles above the Earth could have a devastating effect on communications systems over an area 1500 miles in diameter. (Adapted from John Steinbruner, "Launch under Attack." Copyright 1984 by SCIENTIFIC AMERICAN, Inc. All rights reserved.)

trical and electronic equipment. The effect is similar to that produced in electrical and electronic systems by a nearby lightning bolt, but about 100 times as intense. Two nuclear weapons exploded in space, one over the eastern and one over the western United States, could knock out most of our communications equipment, computers, and other electrical devices and would shut down entire power grids.

Radiation: Delayed

When a nuclear bomb explodes at *high altitude*, the fission products are carried to great heights by the rising fireball and the subsequent mushroom cloud. They are then carried great distances and fall to earth long after the explosion. This is called *global* or *delayed fallout*. If a bomb explodes at low altitude, where the fireball contacts the ground, great quantities of soil are partially vaporized and swept up into the fireball, and they rise with it to great heights. As the fireball cools, however, fission products condense out on the dirt particles, which will begin to fall to the ground. This *early fallout*, which occurs in the first 24 hours, deposits a great deal of radioactive material on the ground, both in the area where the bomb exploded and in a swath downwind from that area.

Although the early fallout could extend 100 miles from the bomb site, most of it will fall within a few miles. This is where it is most serious, because the farther the fallout cloud goes, the more it will spread out, and the less concentrated it will become. Where the radioactive particles will fall, of course, depends critically on the winds, both aloft and at the surface, and on the precipitation. Therefore, the distribution of early fallout may be very haphazard. At Nagasaki, where the bomb site was surrounded by low mountains, much of the early fallout was deposited in a river valley just over the mountains to the east.

The composition of early fallout is somewhat similar to that of a drastic nuclear reactor accident, such as the one at Chernobyl. The hazard is due primarily to short-lived fission products, such as iodine-131, and the decay is very rapid. A lethal dose for a human—450 rads—might be accumulated in the first half hour after the explosion. However, someone entering the area after a week would accumulate only 100 rads during the next year.[4] Therefore, after a week, people could enter the area for short periods of time. After a year, they should be able to begin to repopulate the area, although careful radiation monitoring would be needed until soil, buildings, and debris containing long-lived radioisotopes had been removed.

Nevertheless, even though most of the radioactive exposure will occur in the first week, the *total* radioactivity spread across the countryside will by no means be negligible. A single 10-megaton bomb exploded at the surface would deliver an average lifetime dose of 10 rems to the entire population of an area the size of California. This dose represents a 50% increase over the average lifetime dose from background radiation.

Immediate Effects: Biological

A nuclear war between the U.S. and the Soviet Union might well be all over in a few hours. It could produce 300 million deaths *within the first week*, from blast, heat, and fires. If Europe were involved, the number would be more like 500 million deaths, or $1/10$ of the population of the world.

Metropolitan areas like New York might be hit by 20 or 30 bombs. To make the picture more comprehensible, let us consider the effect of just *a single 1-megaton bomb* exploded at a 1-mile altitude over a large American city, such as Detroit. The physical consequences of such an explosion have been described above. We will now consider the *human* consequences.

Making the reasonable assumption that a firestorm would develop, this single explosion would kill about one million people, or $1/4$ of Detroit's population. Another million—one out of three of the survivors—would be seriously injured. The number of people with third-degree burns would exceed 200,000, which is 100 times the number of intensive-care burn beds in the entire U.S. Because the bomb would be exploded over the city center, presumably during work hours, about $3/4$ of the hospitals would be destroyed, and most of the medical personnel killed. There would be no more than 1 surviving physician per 1000 injured people. Assuming that this physician could find the injured with no loss of time—an improbable scenario—and that only 15 minutes per patient was spent on every aspect of diagnosis and treatment, and that the physician worked an 18-hour day, it would still take him or her 2 weeks to see every patient. Because of debris in the streets and lack of communication systems, one could not expect ambulance service, but even if one did get to a hospital, 80% of the beds would already be occupied by pre-bombing patients. Most of the victims would never get medical care and would die slow and agonizing deaths, without the benefit of narcotics or even bandages.

This is not mere speculation. Of the 300 physicians who lived in Hiroshima, only 28 remained active after the atomic bombing. Nearly all the city's medical supplies were burned in the firestorm, and every hospital except one was completely destroyed.[5]

It must be recognized that such an explosion over the city center would devastate civil services. Radio, TV, and central telephone equipment would be destroyed, so there would be little communication other than by word of mouth. Many radios and TVs would have been knocked out by the electromagnetic pulse. Electric power plants and power lines would be inactive, so there would be no electricity, no lights at night, and no heat from either electricity or gas. Water mains would be ruptured, and no water would come out of the tap. Burned people develop a great thirst, but there would be little safe water for them to drink. All central roads would be destroyed, as well as railroad stations and warehouses, so no food would be delivered to what was left of the supermarkets. Within a week, most survivors would be starving and severely dehydrated.

The psychological trauma would be incredible. Imagine standing, or more likely sitting with your head in your hands, on your burned-out front lawn after having miraculously survived the fire that burned your house down, surveying the devastation around you, knowing or suspecting that most of your loved ones are dead or even now dying painfully, beginning to feel the pain from your skin burns, and knowing that there is no place to go for help. This trauma alone would kill some people. Many injured people would simply go into a state of shock, as they did at Hiroshima. Such shock can be lethal.

This scenario, of course, assumes that there would be no outside help. Is this realistic? Remember that 5000 weapons would have been exploded throughout the country. Suppose you live in the suburbs of Dallas, where you *might* survive. Your city alone would have received at least 4 bombs. Fort Worth would have received the same, because of the Carswell Air Force SAC base located at the city limits. The Dallas-Fort Worth airport would have been hit. Houston, Austin, San Antonio, and Oklahoma City would have been hit, including air force bases near each of these cities, as well as Corpus Christi, a naval base. No major help, either as medical aid from other cities, or as governmental help from central city offices or military bases, would be likely. Assume the improbable event that your car still ran—after the shock to its ignition system from the electromagnetic pulse—and that it had a full tank of gas. You might drive out of the city, assuming you could get past the rubble and avoid the unfortunates who would beg for or demand that you give them a ride. In the country, you might find food, water, and shelter, but you would be competing with thousands of others for the hospitality of the rural residents, who might be reluctant to share their limited supplies, knowing that there would be nothing left after these were gone.

If you survived the initial blast and heat, and *if* you managed to escape from the city, you might then begin to suffer from "acute" or "short-term" radiation effects.

Acute Effects of Radiation

The biological effects of low doses of radiation are very difficult to detect, as we discussed in Chapter 4. In fact, with doses as high as 50 rems, there will usually be no overt symptoms of radiation exposure; we can merely calculate an increased risk of mutation and cancer over many years in a large population.

But the effects of *high doses*—over 50 rems—are a different matter. A whole-body dose of 450 rems will kill 50% of a human population that gets good medical attention, and 600 rems will kill all of those exposed. For anyone receiving over 600 rems, the incidence of various radiation-induced diseases is irrelevant, because the person will die anyway within a couple of weeks. I will therefore discuss what happens to those who receive between 50 and 600 rems. Such exposures were experienced by about 10,000 survivors of Hiroshima and Nagasaki combined, perhaps 1000 people at or near the Chernobyl reactor, and a few victims of various other radiation accidents. Long-term data from Chernobyl are not yet available, so we are left mostly with data from Hiroshima and Nagasaki. I start with the effects produced at the lowest doses and move progressively to more severe effects observed at higher doses. It should be recognized, however, that *at higher doses, all the lower-dose effects will also occur*.

In the lower dose range of **50–200 rems**, the chief symptom is a *lowered white-blood-cell count*. There may be some nausea and fatigue.

The next level, **200–300 rems**, involves a more marked drop in the white-cell count. Such a lowered white-cell count means that the immune system has been compromised, causing a *lowered resistance to disease*. This, in fact, is one of the major dangers of high radiation exposure, especially if, as in a nuclear war, medical attention is not available and disease germs are multiplying, because of broken sewer systems, the release of stored food from refrigerators and markets, and unattended corpses. In this dose range, there will also be marked *nausea and fatigue*, with some diarrhea.

At the next level, **300–450 rems**, there is a characteristic *loss of hair* after a week or so; the beginnings of *damage to the gastrointestinal system*, reflected in strong *diarrhea*; and *damage to all blood cells, as well as the organs, chiefly the bone marrow, that produce them*. In such victims, the red-blood-cell count drops, leading to *anemia*, and the white-blood-cell count drops drastically for about three weeks, then slowly recovers over another month or so.

At the highest doses of radiation that can be survived **(450– 600 rems)**, one sees severe action on the bone marrow and the gastrointestinal system. Thirteen victims of the Chernobyl accident were given bone marrow transplants by Robert P. Gale, a physician at the UCLA Medical Center, and this treatment apparently saved two lives. There will be the beginnings of *effects on the nervous system*, chiefly those nerves that control our internal organs. The manifestations of such damage are nausea, vomiting, and diarrhea, with consequent loss of body fluids, often followed within a week by *fever, apathy, and delirium*. After a nuclear bombing, much of the delirium would be due to the devastating psychological experience of seeing one's world instantly and completely destroyed, and one's family and friends dead or dying. At Hiroshima, people were found following each other in single file along a dirt path through the rubble, instead of walking on the road; they were too dazed to think. People with these symptoms usually die within a few weeks, but some may survive if given expert medical treatment.

As we have noted, the severe immediate effects of radiation are of little consequence for many victims of a nuclear bombing, as the blast and heat kills most of those exposed to high radiation doses. *Relatively* few survivors will have high radiation damage.

Such highly irradiated survivors of the blast and heat will be in trouble. Because radiation suppresses the immune system, and there will be little medical attention, disease will spread rapidly to these weakened persons. Radiation biologists have estimated that, because of such disease, as well as the extreme psychological trauma that victims will experience, *the lethal dose of radiation in a nuclear war would drop from the 450 rems that is normal with good medical care to a value closer to 250 rems.*[7] In other words, during a nuclear war, roughly twice as many people would die of the consequences of a given radiation dose than would die in a peacetime radiation accident.

Induced radioactivity—radioactivity induced by neutron irradiation of the ground and buildings—will be located close to ground zero, so that most of those exposed to it will be killed by the explosion itself, unless they are well shielded from the blast and heat. Most survivors will leave the area quickly. The induced radioactivity is of significance chiefly for people who re-enter the area shortly after the explosion.

Early fallout will be of significance for those downwind of the explosion. Such fallout will occur over a period of up to a day, but it may remain on the ground for weeks, so that it can be inhaled or picked up by shoes or clothing. This ground contamination will also be absorbed by plants, which may then be dangerous to eat. Exposed people who leave the area and wash soon thereafter will not be in great danger, but anyone who remains in a location up to 20 miles downwind may accumulate a high dose in a few days or weeks. It is for this reason that people living within 20 miles of the Chernobyl power-plant explosion were evacuated.

The combination of induced radioactivity and fallout would leave much of the area of the explosion uninhabitable for many months. Without careful cleanup operations, this period could extend to years.

Finally, those who survive all the above early effects of a nuclear bombing then enter the "delayed-effects" scenario, involving global fallout and a phenomenon called *nuclear winter*. These effects are harder to run away from.

14 Nuclear War
II. The Slow Death

A qualitative difference between nuclear war and conventional war lies in the delayed or late effects. Such effects would be a major consequence of a nuclear war, whereas they are quite minor in conventional war. There are two types of late effects of nuclear war: those caused by *lingering radiation*, which induces cancer and sometimes death, and those caused by *changes in the atmosphere*, which alter the climate.

Lingering radiation results from induced radioactivity in the soil at the site of the bombing, and from radioactive fallout. As we noted in the last chapter, induced radioactivity would not be a major health problem, because survivors would quickly vacate the bomb area. However, radioactive fallout would be a major hazard. *Early fallout* downwind of the explosion may extend for 100 miles. It will fall to the ground within a day, but may contaminate the ground with moderately high radioactivity for years. Global or *delayed fallout*, from particles lofted high into the atmosphere by very large weapons, may take months to reach the ground. Because of the wide dispersal of this fallout, it will produce relatively low radioactivity, but fission-product isotopes with long half-lives, like cesium-137 and strontium-90, may be a long-term problem.

In the first week after a nuclear explosion, the major hazard of fallout is the inhalation of radioactive particles. Later, for up to several years, the major hazard is gamma rays emitted by radio-isotopes on and in the ground. The only way to decrease this hazard is to scrape away the soil and either bury it or transport it to some place away from human habitation. Eventually, most of this radioactivity will be washed out of the ground and diluted to innocuous levels. For many years, however, it will create a food hazard. Plants will take up the radioactive atoms, and animals will eat some of these plants. Thus, both meat and vegetables will be affected.

Delayed Effects of Radiation: Hiroshima and Nagasaki

Somatic Effects: Cancer

One of the major delayed effects of radiation is, of course, the induction of cancer. This is caused by mutation in *somatic* cells, which are all of the cells in the body except the germ cells. Cancer may be induced not only by radiation, but also by chemicals in air, water, and food. Because we are exposed in our normal daily life to a wide variety of cancer-producing substances, including background radiation, it is not surprising that 1 out of 3 Americans will eventually contract cancer, and 1 out of 5 will die of it. Most "natural" cancers take a long time to develop, on the order of 25 years; so cancer, like heart disease and stroke, is largely a disease of old age. This is also true of most radiation-induced cancer. Speaking of the A-bomb survivors, epidemiologist Dr. Hiroo Kato, of the Radiation Effects Research Foundation in Hiroshima, said, "In general, radiation-induced solid cancer begins to appear after the age is attained at which the cancer is normally prone to develop (so-called cancer age)."[1] This means that, except for cancers of the blood or bone marrow, like leukemia, people exposed to radiation will not develop a particular cancer any sooner than unexposed people but that, *after the age at which people normally begin to get that cancer*, the exposed people will show a higher incidence of the disease.

Cancer deaths at Hiroshima and Nagasaki have not been high. A model of those who received high doses—adjusted to a dose of 100 rads—showed the risk of leukemia death to be 5 times normal (relative risk of 5 in Figure 35), reaching a peak 5 to 8 years after the bombing. Risk of deaths from multiple myeloma, a cancer of the bone marrow, was 3 times normal. But death risks from other (solid) tumors averaged only 1.4 times normal, most appearing later than 20 years after the bombings. No significant increased risk was observed for cancers of the rectum, gall bladder, pancreas, liver, uterus, prostate, mouth, nose, throat, bone, and brain, or for skin cancer (except melanoma) and malignant lymphoma. Thyroid cancer can usu-

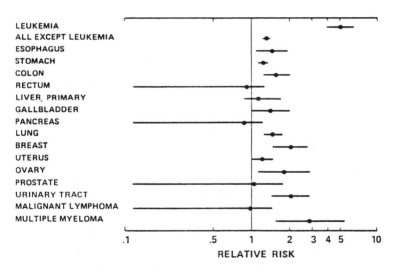

Figure 35. Estimated relative risk of death due to cancer during 1950-1985 among survivors of Hiroshima and Nagasaki, adjusted to a radiation dose of 100 rads. The *relative risk* is the ratio of the number of cancer deaths among irradiated survivors to the number of cancer deaths expected among unexposed survivors. The horizontal bars indicate 90% confidence intervals. Note the logarithmic scale of relative risk on the horizontal axis. (From Shimizu, Kato, and Schull, 1988— see Bibliography. Reprinted by courtesy of Radiation Effects Research Foundation, Hiroshima, Japan.) See also Pierce, 1989—see Bibliography.

ally be cured with radioiodine treatment and is therefore not listed in Figure 35 as a lethal cancer. Considering these facts, we may conclude that *the only lethal cancers whose induction was importantly increased by nuclear bombing were cancers of the blood or bone marrow, such as leukemia and myeloma.*

Among 18,000 survivors of Hiroshima and Nagasaki who received doses of at least 10 rems—and thus might be expected to have a higher risk of cancer—only 325 excess cancer deaths had occurred by 1985.[2] About one-quarter of these were leukemias. Few additional leukemias are likely to appear, but late development of solid tumors may conceivably raise the final number of excess deaths to 600. This is about 3% of these survivors. Thus, *survivors of the atomic bombings who received doses greater than 10 rems have only about a 3% chance of dying from cancer due to the bombings, and most of these cancer victims will die at an advanced age.*

In addition, there may be *radiation dose thresholds*—doses below which there is no effect (see Figure 11)—for the induction of cancers in Hiroshima-Nagasaki victims. Sohei Kondo, of Kinki University in Osaka, described significant dose thresholds for the induction of leukemia and some other tumors.[3] A committee of the U.S. National Academy of Sciences, however, concluded that, among the A-bomb victims, *only* leukemia has shown a sublinear dependence on dose. They did admit, however, that the "epidemiologic data cannot rigorously exclude the existence of a threshold" for other tumors.[4] The question seems unresolved, but it is important, because the existence of a threshold means that there would be *no* cancer induction at very low doses.

Other Somatic Effects

Birth defects among children irradiated *in utero* are caused by somatic effects in the fetus, rather than by genetic effects in the mother. There were many miscarriages after Hiroshima and Nagasaki, and many children were born with various defects.

For 30 mothers pregnant at the time of the bombing, who were within 1.2 miles of ground zero at Nagasaki, 60% of the children (18 cases) either died before or after birth or were severely mentally retarded, compared with 6% in unexposed children.

High numbers of chromosome aberrations—permanent changes in chromosomes that are considered to be a kind of mutation—have been observed in the bombing survivors. Such excessive aberrations were not found in offspring conceived after the bombing.

Germ-Cell Effects

Among those exposed to moderate or high doses of radiation, the *sperm count* will drop and may remain at a lower level for many years. This effect greatly reduces the chance that children will be conceived from irradiated sperm.

But how about the precursor cells that *produce* the sperm? Would they be damaged? Yes, but data obtained with both mice and humans show that they are remarkably resistant. No long-term effects on fertility have been observed among A-bomb survivors.

The human egg is a different story. All of the immature cells, called *primary oocytes*, that will become the eggs that a woman ovulates over her lifetime are *already made at the time of her birth*. Yet, again, data obtained with humans confirm that these primary oocytes are very resistant to radiation damage.

It is thus not totally surprising that Dr. James V. Neel and co-workers, associated with the Radiation Effects Research Foundation in Hiroshima, have found no statistically significant increases in eight different mutational effects in the offspring of survivors of Hiroshima and Nagasaki.[5] These eight effects include untoward pregnancy outcomes; infant deaths, cancers, and abnormal development; and chromosomal damages and other mutations. Similar results are appearing for children born to women exposed at Chernobyl.

However, such effects may still occur in future generations, owing to **recessive mutations**. These are mutations that occur in a gene on only one chromosome of a chromosome pair and that show no effect in that person. However, in future matings, such a defect from one parent may match up with a similar one from the other parent (not necessarily caused by radiation), and the result may be children with the defect controlled by that gene. Radiation damage may thus show up only in some distant future generation. However, the fact that no genetic effects have appeared in the 45 years since the atomic bombings means that *long-term genetic damage appears to be low*. The reason may be a temporary halt in fertility of the fathers (thus disposing of damaged sperm), as well as repair processes in the germ cells of both parents.

In summary, we can make the following generalizations about the effects of exposure to nuclear-bomb radiation: (1) cancer deaths are relatively low, and (2) there are very few genetic abnormalities: children born after the bombing do not show significantly higher levels of such defects.

How can one explain these low effects, for we know that mutations induced by radiation do cause cancer and genetic abnormalities in laboratory animals?

Of those people who received *high doses* of radiation in the atomic bombings, it appears that most were killed by the blast and heat. The fraction of people who get a high dose of radiation and are not killed by other effects depends very much on the particular type of bomb used and the altitude at which it is detonated. At Hiroshima and Nagasaki, this fraction was low. Among those who do survive with such high radiation doses, there is, as noted above, much abortion of fetuses and reduced sperm counts. Thus, much of the damage to future generations is avoided.

For those who received *low doses* of radiation in the atomic bombings, there may be many explanations of the lower-than-expected effects. Biological organisms have evolved over a period of 4 billion years in an environment of background radiation even higher than that we experience today. Radioisotopes with half-lives less than 100 million years have long ago disappeared. Even the isotopes with half-lives of billions of years, such as uranium-238, were much more abundant in earlier epochs. Nature apparently learned to cope with this background radiation in several ways, such as (1) very efficient repair of the DNA by special repair enzymes; (2) the DNA being in a dormant physical state in sperm, which protects it from radiation damage; (3) stimulation of the immune system by low doses of radiation; or (4) a highly selective "killing off," by the organism, of cells that have been seriously damaged. On this last point, Sohei Kondo has suggested that, in certain body tissues, low-level radiation may stimulate the rapid death of precancerous cells, which are then replaced by new, healthy cells.[6] Kondo noted that low radiation doses actually *decrease* some cancer induction in experimental animals. These ideas are not new, but they are still somewhat speculative.

However, one should not be lulled into complacency. Although cancer induction and birth defects were not great at Hiroshima and Nagasaki, the reason is largely that those who got very high levels of radiation were killed by the blast and the heat. Such favorable outcomes would not occur for people far from a detonation who got high doses of radiation from fallout. In an all-out nuclear war, about 30% of the mid-latitude land area of the Northern Hemisphere would be exposed to 500 rems of ionizing radiation—X rays, gamma rays, and neutrons—within a day. Such radiation doses would kill more than half the *healthy* humans exposed in an *undamaged* environment. But it would kill most of those exposed after a nuclear war, because of lack of medical attention, the lowering of the immune response to widespread disease, and psychological trauma.

Furthermore, there will be some *very* long-term effects, chiefly in the form of recessive mutations, which may occur in future generations. These might include reduced resistance to disease, or genetic imperfections leading to impaired functions, such as lowered mental ability or an increased need for certain "vitamins" in food. These mutations would reduce the vigor of the entire human race and could take a thousand years to overcome.

Other mammals, as well as birds, would be killed by doses of about 500 rems. The loss of birds would increase insect populations, which already have the advantage of being very resistant to ionizing radiation. The plagues of insects that we might suffer after the ecological dislocations of a nuclear war would make the biblical plagues seem like minor disturbances.

Plants are generally more resistant to ionizing radiation than mammals. However, evergreen trees are quite susceptible and would be killed in large numbers. Such dead firs and pines would be excellent fuel for forest fires.

Nuclear Winter[7]

Some say the world will end in fire;
Some say in ice.

ROBERT FROST

How miserable one can be on a cold January day! The skies are gray, the wind blows around the house, and the rooms are drafty. We put on sweaters and maybe start a fire in the fireplace. We're glad we don't have to be outside.

But what if we had no electricity or heating fuel? Unless we had an unusually large supply of firewood—and a fireplace— we would be in danger of freezing. We would certainly run the risk of catching a cold or pneumonia. We would have no lights and no way to cook our food except in the fireplace. The house plumbing would freeze, cutting off our supply of water. We would be in very serious trouble, just from these *immediate* consequences of a nuclear war, if it happened during the cold winter months.

But in fact, the story is much worse than this. We now know that the skies would darken for months after a nuclear war, and it would be cold even in July. This phenomenon is called **nuclear**

winter, and it was first seriously proposed and extensively studied in 1983 by Richard P. Turco at R & D Associates, Marina del Rey, California; O. B. Toon, T. P. Ackerman, and J. B. Pollack, NASA Ames Research Center, Moffett Field, California; and Carl Sagan, Cornell University.[8] (This is known as the "TTAPS" paper, from the initials of the authors' last names.) If the immediate consequences of nuclear war seem horrible, the added consideration of nuclear winter makes any planning for such warfare seem insane.

Nuclear winter would be caused mostly by small particles of dust that are raised by nuclear explosions well up into the stratosphere—that region of the atmosphere above about 50,000 feet, where the temperature is very low and the air is quiet, with no clouds and little wind. Large nuclear explosions where the fireball touches the ground will vaporize, melt, and pulverize the surface of the earth, sucking large quantities of condensates and fine dust into a hot column of air that rises into the stratosphere. These particles may remain in the stratosphere for as long as a year. They will blot out the sun, and the surface of the earth will cool.

In addition, bombs that explode in cities or near forests will produce great firestorms, which will lift massive quantities of smoke particles into the atmosphere below the stratosphere. The ignition of oil-storage tanks and coal piles will add to the smoke produced. This smoke will also blot out the sun, although much of it will fall out of the lower atmosphere in a month or so. Up to a third of the smoke, however, may be lofted by the largest firestorms into the stratosphere, to contribute to the long-term dust effects described above.

Consider the scenario for a nuclear war that is both counterforce (directed against military targets) and countervalue (directed against cities), involving 5000 megatons (MT) of explosive force. Such a war would involve using only 30% of the world's nuclear arsenals (Table 6). After such a war, in many parts of the target countries it would be too dark to see for several weeks. Averaged over the whole Northern Hemisphere, the light would be only a small percentage of its normal value, comparable to the light under a dense overcast. If the normal daytime temperature were 56 degrees Fahrenheit (spring or fall temperatures), at the end of a month it would have dropped to *7 degrees below zero Fahrenheit, or -22 degrees Celsius!* After three months, the temperature would be back up to freezing, but it would not get back to normal until a year after the war.

Suppose we had a much smaller and "limited" countervalue nuclear war, involving the explosion of only 100 MT, all targeted on cities. The temperature would *still* drop to 7 degrees below zero within a month. But it would recover more quickly, getting back up to freezing after 2 months, and reaching normal temperatures in about 4 months.

In *countervalue* attacks, bombs would be exploded high in the air, to produce effects over a large area, and the fireball would usually not touch the ground. The result would be firestorms from burning debris in the city, as well as forest fires in the suburbs. This would produce smoke, which would stay mostly in the lower atmosphere and would result in a more-severe, but shorter, nuclear winter. In *counterforce* attacks, however, bombs would be aimed at missile silos and hardened command posts and would be exploded near the ground. Such an attack would produce little fire, but tremendous amounts of dust would be injected into the stratosphere. The result would be an equally severe nuclear winter, and one that would last much longer: after a 5000-MT counterforce war, the temperature would stay below freezing for a year.

An important point to stress is that bombs of yield below 100 kilotons will not inject significant amounts of material into the stratosphere, but the fireball of a 1-megaton weapon will rise almost entirely into the stratosphere. This fact has an important bearing on arms control, as smaller weapons will produce a less-severe and shorter nuclear winter.

What effects would such temperature drops have? To put this situation into perspective, consider the fact that a year-round three-degree Celsius (5° F) temperature drop would eliminate wheat and barley growing in Canada.[9] Much of the world's cereal grain is produced in high northern latitudes, where the growing season of several months is just barely long enough to support a crop. Shortening the growing season by only a week—about 4 days at each end of the season—could preclude growing the crop. Such a slight shortening can be produced by a very small average decrease in temperature.[10]

Virtually all vegetation, including crops and wild plants, in the mid-latitudes (30°–70° latitude) of the Northern Hemisphere would be killed in a nuclear war that occurred just before or during the growing season. If the war occurred after harvest or seeding time, the nuclear winter would preclude a successful crop for the next season. Thus, regardless of when the war occurred, one year's plant production would be lost throughout the mid-latitudes of the Northern Hemisphere.

As a matter of fact, *even if temperatures remained normal*, the production of crops and wild plants would still be greatly reduced, simply because of the lack of sunlight.

Loss of crops means the death of livestock. Together, they mean loss of most food for humans. Ocean fish would still be available, as the oceans would not cool significantly because of their tremendous heat content. However, the green marine plants, which grow by photosynthesis using the light of the sun, lie at the bottom of the food chain, and they would be decimated, which would cause a drop in ocean fish populations. Most streams and lakes would freeze; fish in the others would die from the lack of green aquatic plants. Game animals would die from cold and lack of food. What crops or animals did survive a nuclear winter would have to be harvested or killed by men and machines, and transported to market. But healthy men, machines, and transport would not be available.

In contemplating nuclear war, we are talking disaster, even if we consider *only* the initial effects of the explosions. Add to these the freezing temperatures and the lack of food, and we have a picture of wholesale death—not only in the early weeks after a war but extending for years into the future.

Other Delayed Effects

Disease

The large number of unattended corpses, as well as rotting food from useless refrigerators and warehouses, and sewage from broken disposal systems, would quickly swell the numbers of rats and other rodents. Many rodents would survive, because they were sheltered underground. Insects, notoriously resistant to radiation, would flourish as well. The proliferation of these pests would result in the rapid spread of disease. Human survivors, weakened by burns, irradiation, and other severe traumas, would be vulnerable. Furthermore, once infected, they would have few medicines with which to fight the disease; they would be hungry, and at any other season except midsummer, they would be very cold. Epidemic disease would sweep the land.

Global Effects

The most severe effects of a nuclear war would first be felt by the target nations, presumably in the Northern Hemisphere. Within a week, however, the nuclear winter effects would spread throughout a band encircling the Northern Hemisphere from about 30 to 70 degrees latitude, an area that includes half the world's population. In this band, the light level would be only two-tenths of one percent of normal. The smoke and dust would soon spread over the whole Northern Hemisphere, resulting in an average light level only 5% of the normal level. The entire hemisphere would experience severe nuclear winter.

Would the Southern Hemisphere be spared? Unfortunately not. Normally, the air in the two hemispheres of the Earth is kept well-separated, because hot air at the equator rises into the cold stratosphere (-67°F) and spreads out poleward, falling back to lower levels in the tropics. The dust causing nuclear winter would absorb sunlight in the stratosphere and heat it up, at the same time that air near the surface was being shaded and therefore cooling rapidly. Thus, the normal warm-surface–cold-

Figure 36. Nuclear winter. *Top*: Looking down on North Pole. Nuclear explosions over cities and forests would light huge, uncontrollable fires, which can be seen burning on the dark side of the planet. *Bottom*: Looking in at Equator. Explosions near the ground would inject a tremendous amount of dust into the stratosphere, where it would persist as long as a year and would spread all over the globe. (From Laura Tangley, 1984, "After nuclear war—A nuclear winter," *BioScience* **34** (January), 6–9. Artwork copyright 1983 by Jon Lomberg.)

stratosphere situation in the tropics would change to one of cold-surface–warm-stratosphere. This new situation would cause stratospheric air to flow toward the equator and to spill over from the Northern into the Southern Hemisphere. Thus, nuclear winter would spread throughout the world (Figure 36). On a global basis, the 200 million tons of smoke of a nuclear winter would lower the *average* light to 25% of its normal level. Such a loss of sunlight would still be enough to produce severe effects.

The spread of stratospheric dust throughout the world means that radioactive fallout would also extend worldwide. Most of the survivors in the mid-latitudes of the Northern Hemisphere would die within a year of radiation exposure, or of disease resulting from their radiation-damaged immune systems. People elsewhere in the world would probably not get lethal doses of radiation, but many would get as much as 100 rems and very few less than 10 rems over the years following the holocaust. These doses spread over a year or two might not be serious for individuals, but the long-term consequences for the world population, because of recessive mutation, could be profound.

Another global effect would work on wild animal populations. Both humans and animals are protected from the dangerous high-energy ultraviolet rays of the sun by a layer of ozone (a chemically active form of oxygen) in the stratosphere. Widespread nuclear explosions would inject nitrogen oxides into the stratosphere, which would in turn cause severe *ozone depletion* for years after the war. The resulting flow of high-energy ultraviolet light down to the Earth's surface, after the nuclear winter was over, would damage the eyes of both humans and animals. The humans could wear sunglasses or go indoors, but the wild animals would be blinded and would die—unable to catch their prey.

A year or two after the war, the nuclear winter will largely have disappeared. But some atmospheric changes, like damage to the ozone layer, may last much longer. The ecological dislocations and the disruption of human civilization could last for centuries.

Figure 37 summarizes the major effects of a nuclear war.

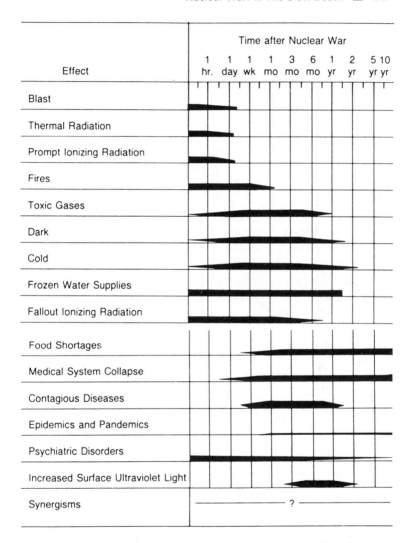

Figure 37. A schematic presentation of the time scale for many of the effects of a 5000-megaton nuclear war. The thickness of the bars represents the severity of the effect. Note that the time scale is greatly compressed on the right-hand side. (From Carl Sagan, in Ehrlich, Sagan, Kennedy, and Roberts, 1984, *The Cold and the Dark*—see Bibliography. Copyright 1984 by Carl Sagan. All rights reserved. Reprinted by permission of Carl Sagan and W. W. Norton & Co., Inc.)

The End of the World?

What I have described so far is what some would consider a fairly conservative estimate of what would happen in a less-than-full-scale nuclear war (5000 MT). The World Health Organization has estimated that a large-scale nuclear war would immediately kill *at least a billion people*, and that at least another billion would be so seriously injured that they would die within a few months. Within a year, another billion would probably die of starvation and disease. These numbers add up to over half the population of the world. Most of the remaining people would slowly starve to death.[11]

The global nuclear winter would stop photosynthesis (the process by which green plants make food) for at least three months. Weather patterns would be vastly altered. The monsoon rains, driven by seasonal winds, are of critical importance to subtropical ecosystems and agriculture, and are the main source of water in these regions. The atmospheric disruptions of nuclear war would destroy this weather pattern. What rainfall did occur would probably be at sea or along coastlines. Much of Western Africa, India, Southeast Asia, China, and Japan would suffer prolonged drought. Widespread fires and the lack of fertilizers and pesticides would add to the disaster.

Would *anyone* survive? The answer is yes. Small groups of people here and there, in tropical and subtropical regions, would manage to eke out an existence. Among these survivors, there would be some radiation-induced cancer and genetic abnormalities. Disease would be rampant for many decades. The total collapse of civilization and organized agriculture would isolate the surviving groups, forcing them into self-sufficiency.

A reasonable estimate is that life would be turned back to the level where it was 10,000 years ago. Of course, it would advance more rapidly than it has historically, because some few survivors would have been well-educated and would have a deep knowledge of civilization. These would become "sages," whose existence would accelerate the recovery of society. Still, it would probably take many centuries to get back to where humankind was before The War.

The concept of nuclear winter is hard for many to accept. This is partly because the idea is based on speculation and partly because of psychological resistance to any theory predicting the end of civilization. Many people have questioned this scenario, and among them are some respected scientists.

Two questions may arise. First, if only 100 MT can produce a severe nuclear winter, why didn't all the atmospheric testing of nuclear weapons done by the United States and the Soviet Union in the 1950s and 1960s produce such effects? Because (1) those tests were spread out over many years, and (2) many of the large tests were done near the ocean surface and produced neither fires nor large columns of dust; the land tests were done in deserts and did not produce fires.

One must remember that nuclear winter is largely a *cumulative* effect. A single nuclear explosion will not do it. But a 100-MT war, involving perhaps 500 explosions, each of 200-kiloton yield, all occurring on the same day, represents roughly the lower limit of what *will* do it. Such a war would involve less than 1% of the world's nuclear arsenals.

The second question is one of mistrust. All of these estimated nuclear winter scenarios are just that: estimates of what would probably happen. What evidence do we have that it *will* happen? Actually, there is quite a bit of supporting evidence. Consider the explosion of the Tambora volcano in Indonesia in 1815. This was the probable cause of an average *world* temperature decline that year of about 1° C, due to obscuration of the sun from fine dust raised into the stratosphere. The next year was known in Europe as "the year without a summer," and that winter in America as "eighteen-hundred-and-froze-to-death." The volcanic explosion of Krakatoa, an island between Java and Sumatra, in 1883, resulted in gorgeous sunsets in England for several years after, due to suspended dust in the stratosphere. Recent observations show that smoke traveling for a few days from fires in British Columbia to the U.S. Midwest lowered daytime temperatures 2–4° C, and that smoke from forest fires in northern California in 1987, trapped by an atmospheric inversion, lowered temperatures by 20° C in the Klamath River Canyon for more than two weeks.[12] So there is hard evidence that something like a nuclear winter could occur.

A study in 1987 by Stephen H. Schneider and his co-workers, at the National Center for Atmospheric Research of the University of Colorado, suggested that the nuclear winter would not be nearly as bad as predicted by the TTAPS study.[13] This report was centered on the idea that the oceans would remain warm, and that this would help to keep the fringes of the continents warm. This possibility was recognized, but not dealt with, by the earlier investigators. Schneider and his co-workers concluded that, in a midsummer war, there would indeed be some freezing, but that it would be spotty and recovery would be rapid. They claimed that the worst-case temperature drops of the earlier study were overstated by at least a factor of 2. However, even they were compelled to say:

> But the climatic effects might nonetheless be calamitous, and they would extend the impact of the war to billions of people who live far from the blast zones. . . the earth's biota can be highly sensitive even to small climatic disturbances.

A 1990 article in *Science* magazine[14] by the original TTAPS proponents of the nuclear winter scenario summarizes all of the nuclear winter research to date, including studies by the U.S. Office of Science and Technology Policy, the U.S. National Academy of Sciences, the International Council of Scientific Unions, the World Meteorological Organization, the Scientific Committee on Problems of the Environment, and the United Nations. They all agreed that "the widespread environmental effects of nuclear war could threaten most of the human population." The report concludes that, although many of the factors that went into the TTAPS model have been revised, most of them cancel each other out, the result being effects not greatly different from those originally proposed.

One should not be deluded by scientific disagreements over details into thinking that nuclear war would be less horrible than is generally envisaged. Even the relatively puny conventional bombs of World War II destroyed most of Hamburg, Dresden, and Tokyo and killed a large fraction of the people living in those cities. There is no serious doubt that an all-out nuclear war will destroy civilization.

15 ⚛ Myth II: You Can't Trust the Russians

It is abundantly clear that a nuclear war would be catastrophic. Therefore, it is logical that we should minimize the possibility of ever entering one.

Virtually everyone, from the President on down, agrees that an excellent way to achieve this goal would be to reduce nuclear weapons, and perhaps even to eliminate them. To this end, the U.S. has always tried to reach satisfactory nuclear-weapons treaties with the USSR. But this goal has often been thwarted by the almost universal American insistance that the Russians cannot be trusted to keep a treaty. This specter of constant Soviet violations of treaties has been raised over and over again as an excuse for increasing our nuclear armaments.

With the present disintegration of the Soviet empire, it may seem that we no longer need to be concerned with this matter. But, there are several reasons why we must still be on guard against nuclear war: (1) neither the U.S. nor the USSR has yet significantly decreased its nuclear arsenals; (2) we do not know how the Soviet Union will evolve politically—Russia, the major component of the USSR, has a history of authoritarian government and mistrust of the West; and (3) other powers, such as China, could become the "Russian bear" of the future.

Consequently, we must maintain our guard as long as the present frightfully large nuclear arsenals exist. But, at the same time, we must resist paranoia. It was sheer paranoia that led us to the excesses of McCarthyism in the 1950s, and it was sheer paranoia that led us to build a nuclear arsenal at least 10 times as large as was needed for deterrence of a nuclear war. In order to arrive at a more rational position, it is important for us to examine how we have reacted to the processes of nuclear treaty-making and treaty-keeping with the USSR. Such a retrospective will also help us to negotiate future treaties aimed at drastic reductions in nuclear arsenals.

The present chapter deals with three aspects of the issue of nuclear treaties:

1. What is the history of our treaties with the USSR?
2. How can we verify that the Soviets, or any other nuclear power, such as China, Pakistan, or Iraq, will not cheat on treaties in the future?
3. Have the Soviets cheated on treaties in the past?

No nuclear-weapons treaties were signed in the period 1980–1986. This chapter deals with the earlier treaties, signed before 1980; the recent ones are discussed in Chapter 17.

History of Treaties (1959–1980)

Not many people are aware that *seven nuclear treaties* involving the U.S. and the USSR were signed and in effect before 1973 (Table 8). Since then, one more has gone into effect, and three others have been signed but not ratified by the U.S. I deal first with the treaties in force; there has been no question of compliance with these treaties.

Treaties in Force

The first treaty was the *Antarctic Treaty*, signed by President Eisenhower in 1959. It prohibits the presence of *any military weapons*, nuclear or otherwise, in Antarctica and calls for on-site

Table 8. Nuclear Arms Control Treaties[a]

		Year in force
Treaties in force		
M	Antarctic Treaty	1961
M	Limited Nuclear Test Ban Treaty	1963
M	Outer Space Treaty	1967
M	Latin America Nuclear Free Zone Treaty	1968
M	Non-Proliferation Treaty	1970
M	Seabed Treaty	1972
B	SALT I—ABM Treaty	1972
B	Intermediate-range Nuclear Forces Treaty	1987
Treaties signed but not ratified by the U.S.		
B	Threshold Test Ban Treaty	1974
B	Peaceful Nuclear Explosions Treaty	1976
B	SALT II Treaty	1979
Treaties not signed		
M	Comprehensive Nuclear Test Ban Treaty	
	Under discussion 1955–1980; suspended by the U.S.	
B	Antisatellite (ASAT) Treaty	
	Under discussion 1977–1980; suspended by the U.S.	
B	START Treaty	
	Under negotiation since 1981	

[a]Capital letters at the left indicate whether an agreement is multilateral (M) or bilateral (B) between the U.S. and the USSR.

inspections. It was signed by all the countries claiming interests in Antarctica and has been adhered to by all parties. Treaties such as this are important, because they restrict the areas where nuclear weapons may be used.

The *Limited Test Ban Treaty* has been one of the most successful and important nuclear treaties. Its history is worth reviewing in some detail, because it is a model of how the U.S. and the USSR can cooperate if they want to. In the 1950s, as part of the development of their nuclear arsenals, both the U.S. and the Soviet Union conducted many nuclear tests in the atmosphere (Figure 38). However, by the end of the decade, public concern over these tests was rising rapidly. Japanese fishermen had been heavily

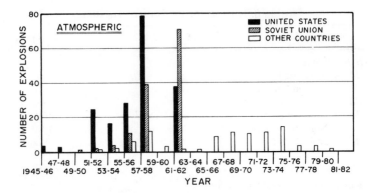

Figure 38. Numbers of atmospheric tests of nuclear devices by the United States, the Soviet Union, and other countries from 1945 to 1982. There have been very few atmospheric tests since 1980. (Data from *World Armaments and Disarmament Yearbooks*, 1976 and 1984, Stockholm International Peace Research Institute (SIPRI); reported in *Nucleus*, Spring 1985, Union of Concerned Scientists—see Bibliography.)

irradiated by fallout from U.S. tests at Bikini Atoll in the South Pacific. Levels of the radioisotope carbon-14 were rising in the atmosphere. Strontium-90 was appearing in milk all over the mid-latitudes of the Northern Hemisphere and was showing up in the bones of children. Consequently, in October 1958, President Eisenhower announced a unilateral moratorium on atmospheric testing. Premier Khrushchev said the Soviet Union would do the same. Eisenhower also called for negotiations on a *comprehensive* test-ban treaty, which would ban all tests of nuclear weapons anywhere.

The U.S. Defense Department and the Atomic Energy Commission both opposed this moratorium, on the grounds that it would be difficult to verify compliance by the Soviets. Such mistrust of the Russians had its impact, for only a year later, Eisenhower suddenly declared the U.S. no longer bound by the moratorium. Khrushchev promptly said the same for the USSR. Then events took an ominous turn. Early in 1960, the French tested their first nuclear device. In May of that year, the Soviets shot down an American U-2 spy plane flying over Russia. In April 1961, the Americans launched their abortive attack on Cuba at the Bay of Pigs, and that summer the Soviets erected the Berlin Wall. Finally, on 1 September 1961, the Soviets resumed atmospheric nuclear testing. Two weeks later, the U.S. followed suit. Then, in October 1962, came the Cuban missile crisis, the closest the two superpowers had ever come to a nuclear war.

Then something happened. Perhaps both sides became frightened by the deterioration of relations and worried about the increased poisoning of the environment by atmospheric testing. In June 1963, President Kennedy declared another unilateral U.S. moratorium on atmospheric testing, and in August 1963, the U.S., the USSR, and Great Britain signed the Limited Test Ban Treaty, which *banned all nuclear testing in the atmosphere, in space, and under water* but permitted underground testing. The treaty was negotiated in only 12 days, largely because of the herculean efforts of Kennedy's special envoy to Moscow, Averill Harriman. This treaty has now been signed by 112 nations but, significantly, not by France or China, both of which continue, although infrequently, to test nuclear weapons in the atmosphere. This treaty has never been violated by those who signed it; it stands as a model of Soviet-American cooperation.

The Limited Test Ban Treaty almost became a comprehensive test ban, which would have included a ban on underground testing. For that provision, the U.S. wanted 7 on-site inspections per year, but the Soviets would permit only 3, so there was no accord. On such a narrow thread hung the possible elimination of the entire future nuclear arms race—but the thread broke.

Other treaties followed in rapid succession. The *Outer Space Treaty* in 1967 banned *all weapons of mass destruction*, not just nuclear weapons, from outer space. In 1968, the *Latin American Nuclear Free Zone Treaty*, or Treaty of Tlatelolco, banned nuclear weapons from Latin America. It was not signed by Cuba and Guyana, and is not in force in Chile, Argentina, or Brazil. As Argentina and Brazil clearly have the potential to develop nuclear weapons, this treaty is seriously limited. The *Non-Proliferation Treaty* of 1970 bars nuclear nations from helping others to build nuclear weapons but encourages them to help others develop nuclear power. It has been signed by 134 countries, but *not* by China, France, Israel, South Africa, India, Pakistan, Brazil, and Argentina. It is thus, like the Latin American treaty, seriously limited. The *Seabed Treaty* of 1972 bans *weapons of mass destruction* from the ocean floor. The first *Strategic Arms Limitation Talks (SALT I)* agreements were signed in 1972. They limit the numbers and types of strategic weapons of both superpowers. This treaty was adhered to by both nations and effectively expired when SALT II was negotiated. SALT I included an *Antiballistic Missile (ABM) Treaty*, which generally prohibits ABM weapons—missiles or rockets that can shoot down incoming ballistic missiles—but permits an antiballistic missile complex around *one site only* in each nation. The Soviets chose to place theirs around Moscow: it includes radars and about 100 missiles. This system is old and is considered of limited effectiveness. We set ours up around our Minuteman ICBM silo complex at Grand Forks, North Dakota, but unilaterally deactivated it in 1975.

Treaties Signed but Not Ratified

Three important treaties have been signed by the USSR and the U.S. but have not been ratified by the U.S. Senate. One is the *Peaceful Nuclear Explosions Treaty*, signed by President Ford in 1976, but not ratified by the Senate. This harmless treaty simply requires on-site observation—by the other country—of all nuclear explosions conducted for peaceful purposes, such as the excavation of lakes or channels. It is, however, not of great importance, as neither the U.S. nor the USSR have done much of this sort of thing, contrary to the great expectations of the Eisenhower Administration for such useful applications.

The Limited Test Ban Treaty of 1963 allowed underground explosions, but both Eisenhower and Kennedy hoped that eventually a comprehensive test ban could be negotiated. The *Thresh-*

old Test Ban Treaty was a step in this direction. It was signed by President Nixon in 1974 but was never ratified by the U.S. Senate. This is one of the most important treaties we have. It limits underground explosions to 150 kilotons, thus making it impossible to test megaton bombs. It turns out, however, that this is not as great a restriction as one might think because, by the mid-1970s, the superpowers were beginning to recognize that megaton bombs were wasteful anyway. Several smaller bombs would do more damage to a city than one big one, and the delivery vehicle for a smaller bomb is lighter and simpler. Furthermore, there are ways of testing the probable performance of a megaton bomb by using "scaled-down" tests (remember that 150 kilotons is still 10 times the power of the Hiroshima bomb). The U.S. and the USSR both claim that they are honoring the Threshold Test Ban Treaty. However, even though the treaty was not ratified by the U.S., the Reagan administration constantly charged the USSR with violations.

This treaty also calls for on-site observations, which are now being undertaken. However, on-site inspection is not really necessary, for it is not hard to detect a 150-kiloton underground nuclear explosion from outside the country, as we shall see.

The third unratified treaty is the *Strategic Arms Limitations Talks (SALT II)* agreement, an extension of SALT I. This was signed by President Ford in 1976, but many of the provisions remained to be worked out. Several political events, including the Soviet invasion of Afghanistan, led the Carter administration to break off talks. The Reagan administration declared the treaty "fatally flawed" and refused further negotiations under its umbrella. Nevertheless, both sides have generally adhered to the Salt II limitations, which involve the maximum numbers and types of ICBMs permitted on each side.

Treaties Not Signed

The *Antisatellite (ASAT) Treaty*, which would have banned the development of antisatellite weapons—missiles that can shoot down military satellites—was never signed and was abandoned by President Reagan in 1980. However, Congress in 1985 banned further ASAT tests by the U.S., contingent on Soviet cooperation, which appears to have been forthcoming. A *Comprehensive Nuclear Test Ban Treaty*, banning the testing of *any* nuclear weapons, which was advocated by all American administrations prior to that of Reagan, was approved by the U.S. House of Representatives in 1986, but went no further. Such a treaty has never been seriously considered by present administrations, partly because it would limit development of the Strategic Defense Initiative (Star Wars).

We defer discussion of the Intermediate-range Nuclear Forces (INF) Treaty and the Strategic Arms Reduction Talks (START) until Chapter 17.

General

When we look at the history of nuclear weapons agreements, certain facts stand out:

1. The Soviets, under Premier Kosygin in 1972, were the first to claim that anti-ballistic-missile (ABM) weapons are purely defensive. But Defense Secretary Robert McNamara argued at that time that they are not, as *any defense is a good offense*. Today, Mr. McNamara opposes Star Wars on those same grounds. The U.S. government has since done a complete about-face, now claiming that the Star Wars system is purely defensive.

2. Some treaties calling for on-site inspections have been proposed by the Soviets and rejected by the Americans.

3. Three treaties have been signed by both sides, but not ratified by the U.S.

4. The Reagan administration terminated discussions on two treaties (SALT II and ASAT). No other administration has ever terminated treaty discussions, although some have temporarily suspended them.

It appears that we have been no more cooperative than the Soviets in proposing and agreeing to nuclear weapons treaties.

Verification of Treaty Compliance

Many of the treaties that we have made with the Soviet Union concern the types and numbers of weapons that each country has. The easiest way to learn if the other country is adhering to treaty provisions would be to have *honest reports* from the "enemy" of its capabilities, supplemented by *on-site inspection*. Both of these sources of information are feasible. Both we and the Soviets release information concerning numbers of missiles, missile types, numbers of nuclear subs, and so on. A great deal of this information is in the public domain. However, both sides suspect the other of hiding important information, especially on the latest developments, and there is little doubt that such secrets do exist. With regard to on-site inspection, our examination of the history of treaties shows that this has often been proposed but was never actually done before 1987, when experiments in on-site inspection began in both the U.S. and the USSR. They have worked well.

Both sides feel compelled to obtain as much information as possible by what amounts to spying. This includes the classic spying techniques, that is, the gathering of data by spies. But another whole area of "modern spying" involves the observation of enemy activities using sophisticated technologies that operate from outside the borders of the other nation, including observations from space. This kind of observation is called **national technical means** of verification. These techniques are remarkably effective. They include methods using *electromagnetic radiation* (photography, radio, and radar—see Figure 5) and methods that measure *mass vibration* (sonar and seismology).

Three types of information can be obtained by these methods: (1) photography and radar images of *missile launch complexes*, be they stationary, on ships, or on mobile land vehicles; (2) sonar and seismic observation of *nuclear weapon tests*—a great deal can be learned about weapon yields in this way; and (3) photographic observation of *missile tests*. We will now examine each of these techniques in detail.

Photography

We now have the ability to take photographs from a satellite that have a 3-inch resolution! This is comparable to looking at a football game from a seat high in the stands. We can take *slant photographs* from high-flying aircraft that can look 250 miles into the enemy country. This technique is especially useful for the observation of ports and their ships; submarines can be readily identified. The chief disadvantage of photography is that it doesn't work through clouds or at night. Infrared photography—using film that is sensitive to infrared radiation, which will detect sources of heat at night—somewhat alleviates this problem.

Radio

The use of radio primarily involves listening to messages associated with weapons tests, such as the monitoring of **telemetry** information: information that is transmitted by a missile to ground stations.

Radar

The computer storage and integration of data from radar images taken by aircraft and satellites can provide resolution almost as good as that of photography. Aerial or space radar also gives excellent information on the heights of objects and terrain features. The great virtue of radar is that it easily penetrates clouds and can be used at night.

Radar is also used from the ground to detect satellites or incoming missiles. A new development, called **phased-array radar**, moves radar beams electronically, instead of moving the antenna itself, and thus can rove across a large area almost instantaneously. The largest phased-array radars have the incredible ability to detect hundreds of objects the size of a basketball that are reentering the atmosphere 1000 miles away. The construction of a large phased-array radar at Krasnoyarsk, in the interior of the Soviet Union, violated the ABM treaty, as we discuss below.

Sonar

Water is an excellent conductor of sound. Therefore, it is easy to detect, locate, and measure underwater nuclear explosions, even from a great distance.

Seismology

Like water, the earth is also a very good conductor of sound. However, the complexities of geological structures make the interpretation of signals difficult. Nevertheless, the size and the location of earthquakes can be determined with remarkable precision. Underground nuclear explosions can also be detected by seismic monitors. The "signature" of the nuclear explosion is sufficiently different from that of an earthquake that a clear distinction can be made between the two (Figure 39).[1]

Boreholes about 300 feet deep can house seismometers whose output can be automatically telemetered to a satellite. Recent studies have indicated that as few as 25 such *unmanned* stations within the Soviet Union, plus about 15 stations outside the country, will allow the unequivocal identification of any nuclear blast greater than 1 kiloton—even if the blast were timed to coincide with an earthquake or was set off in a large cavern to muffle the seismic waves.[2] Such stations can be made tamperproof: any interference or tampering with a station immediately sends a signal to the satellite. Consequently, *underground nuclear explosions of any size can now be accurately detected, even without on-site personnel.* The development of this capability represents a great advance in the verification of treaty compliance.

In summary, the verification of compliance with treaty provisions is now an advanced science, and a highly reliable one. We can detect nuclear explosions and ballistic-missile

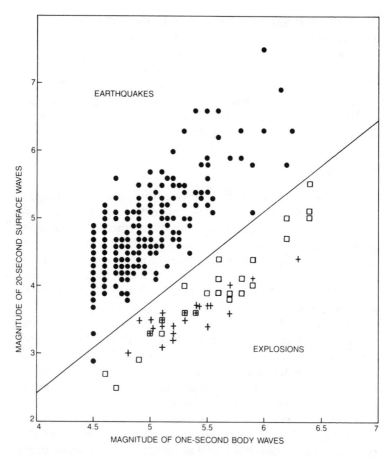

Figure 39. A clear distinction between earthquakes and explosions is evident in this plot of the magnitude of 20-second surface waves against that of 1-second body waves. The 383 earthquakes represented by the black dots were compiled from a set of all the earthquakes recorded worldwide in a 6-month period that had a body-wave magnitude of 4.5 or more and a focal depth of less than 20 miles. The squares represent underground explosions in the U.S., and the crosses represent those in the USSR. (From Lynn R. Sykes and Jack F. Evernden, "The verification of a comprehensive nuclear test ban." Copyright 1982 by SCIENTIFIC AMERICAN, Inc. All rights reserved.)

launches with near certainty. We can also identify stationary missile launch sites with great confidence. The one weakness that remains is the detection of *mobile missile launchers* and *non-ballistic-missile launches*. A small missile launcher can be concealed in a railroad boxcar—we cannot detect this. And it is almost impossible to detect either the launch or the flight of cruise missiles, which hug the ground. Nor can nuclear artillery be easily distinguished from conventional artillery. In short, we can detect *strategic* nuclear weapons with great certainty, but we generally cannot detect *tactical* nuclear weapons.

Have the Soviets Cheated?

The majority of American people have consistently favored restricting nuclear weapons through arms control agreements. Just as consistently, they have said they expect the Soviets to cheat.[3]

RICHARD A. SCRIBNER, Staff Director
Committee on Science, Arms Control, and National Security
American Association for the Advancement of Science

Most of the accusations of Soviet cheating on nuclear treaties were made during the Reagan administration. How true are these allegations? They involve only three treaties.

Threshold Test Ban Violations

The Threshold Test Ban Treaty, *which we have not ratified*, limits underground nuclear explosions to a yield of 150 kilotons. We had some seismic evidence in the 1980s of Soviet tests up to 170 kilotons. However, it is now recognized that any analysis of the seismic pressure waves from underground nuclear weapon tests must take into account the special geological characteristics of each test site. Our failure to recognize this in the past can account for *all* the purported Soviet violations.

There is another very important point with regard to this kind of violation. Suppose that the Soviets had in fact exploded weapons of yield up to 170 kilotons. This is only 13% above the prescribed limit—such a "violation" is of little significance. The object of the treaty is to prohibit very large explosions, of the megaton type. Concern about tests that have only slightly exceeded the limits are little more than quibbling.

Conclusion: The Soviets have not violated the Threshold Test Ban Treaty.

SALT II Violations

The SALT II treaty, *which we have not ratified*, permits the testing and deployment of only one new ICBM by each country. Ours has been the MX missile, which carries 10 warheads. The Soviet's has been the SS-24, a comparable missile, also with 10 warheads (Table 5).

The U.S. claims, however, that the Soviet single-warhead SS-25 is an entirely new missile. The Soviets claim that it is derived from a previous missile (the SS-13) and falls within the 5% modification of earlier missiles permitted by SALT II. Clearly, this is a debatable point. When questioned about our claim of a Soviet violation, President Reagan admitted, on 22 April 1983, that "It is difficult to establish, and have hard and fast evidence, that a treaty has been violated."[4]

The SALT II agreement set a ceiling of 2250 strategic missiles on each side, and also placed limits on MIRVed launchers. The Soviet forces were greater than this at the time the treaty was signed and were never reduced to the required levels. However, by June 1986, the Soviets had removed from operation, or had dismantled, more than 1300 missile launchers, 45 bombers, and 21 submarines.

SALT II permits the encoding of telemetry of missile tests. When the U.S. complained about *excessive* encoding by the Soviets, the latter asked for details, but we refused to document our claims, because we felt that this would reveal too much about our monitoring capabilities!

Conclusion: The only clear Soviet violation of SALT II is in the total numbers of certain missiles, and the excess in numbers is small. However, the Soviets have said that they would comply *if we ratified the treaty*. But the U.S. Senate has consistently failed to do so.

ABM Violations

The ABM treaty permits early-warning radars *at the periphery* of a nation, with the antennas *facing outward*. But it prohibits any other *inland* radars that could be used for ABM purposes (i.e., the tracking of incoming warheads and the direction of interceptors). The Soviets built a phased-array radar at Krasnoyarsk, in central Siberia, that was thus clearly in violation of the ABM treaty, although its frequency of 200 megahertz was ideal for tracking incoming missiles, and was *not well suited to the direction of antimissile battles*. Furthermore, although the Krasnoyarsk radar was in the center of the Soviet Union, it was actually north of Mongolia and at the periphery of the populated part of the USSR, being about as far from Moscow as our phased-array radar at Thule, Greenland, is from New York. The Krasnoyarsk radar clearly filled a gap in the early-warning radar coverage of the USSR.

Conclusion: A group of arms-control experts affiliated with Stanford University have stated that, although the Reagan administration reported 18 different treaty violations by the Soviets, it found only one "actual violation": the Krasnoyarsk radar, which it termed "a technical violation" of the ABM treaty with "little military significance by itself."[5] The Soviets are now, in fact, dismantling the radar.

American Violations

We have not mentioned possible U.S. violations. The Soviets claim that we have violated the ABM treaty by our own phased-array radar at Shemya Island in the Aleutians, which they claim can be used in an ABM mode, much as we once claimed about the Krasnoyarsk radar. They also claimed that we violated SALT II by our deployment of Pershing II and cruise missiles in Europe that could reach the Soviet Union. These missiles have now been removed, under the INF treaty, signed by President Reagan and General Secretary Gorbachev in 1987.

The Reagan administration announced in May 1986 that the U.S. would no longer be bound by SALT II treaty limits. The U.S. then exceeded those limits in November 1986 by commissioning a nuclear submarine in excess of the limit, and by deployment of a B-52 bomber, equipped with cruise missiles, in excess of the number of B-52's permitted by the treaty.[6]

In the proposed deployment of Star Wars devices, the ABM treaty will be violated in general. The Outer Space Treaty will also be violated by the presence in space of nuclear weapons, to be placed there in order to produce an X-ray laser beam that could shoot down incoming ICBMs.

In summary, the purported Soviet violations of nuclear treaties, with the smug assumption that we ourselves never violated such treaties, appears to have been nothing more than a "red herring." But this is a very dangerous game. America flew in the face of rationality and exposed the world to possible nuclear annihilation by its continued insistence that you can't trust the Russians.

16 Nuclear Confrontation

The Arms Race

On 16 July 1945, American scientists produced the first nuclear explosion in history, the "Trinity" test, near Alamogordo, New Mexico. J. Robert Oppenheimer, director of the Los Alamos laboratory that built the bomb, upon observing this terrifying explosion, was moved to quote from the *Bhagavad-Gita*, the Hindu holy poem: "Now I am become death, the destroyer of worlds." A few weeks later, the U.S. exploded over Hiroshima and Nagasaki the only nuclear bombs ever used in war. The entire central regions of both cities were utterly destroyed.

In 1949, the USSR exploded its first nuclear bomb, and so the nuclear arms race began. Many thoughtful people at that time dreaded such a race and cautioned against it. Oppenheimer himself opposed the development of a hydrogen bomb simply because such additional explosive force was not necessary. But the race got completely out of hand, and we must now live with the reality of 50,000 nuclear warheads. The world has about 100 nuclear submarines, the most modern of which have 1000 times the destructive power of the Hiroshima bomb (Figure 40). Yet, nuclear submarines represent only one of a *triad* of deadly weapons systems that includes bombers and land-based ballistic mis-

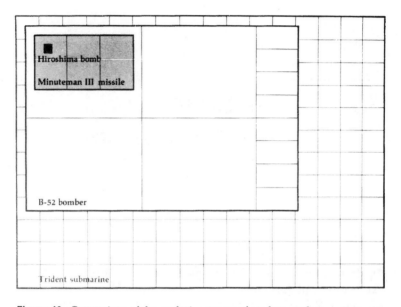

Figure 40. Comparison of the explosive power of modern nuclear weapon systems with that of the Hiroshima bomb, which had a yield of 15 kilotons (kT). *The areas of the boxes are proportional to weapon yield.* The Minuteman III missile carries three independently targetable 335-kT warheads. A B-52 bomber typically carries four 2-megaton bombs (large squares) and eight short-range 190-kT attack missiles; as of 1985, about one-third of our B-52's each carried twelve 200-kT cruise missiles. A Trident submarine carries 24 Trident I missiles, each with eight 100-kT warheads, for a total of 192 independently targetable warheads (1 square = 1 warhead). (*Nuclear Weapons Databook*, 1985, Cambridge, MA: Ballinger. Reported in *Nucleus* **7**, Summer 1985. Courtesy of Union of Concerned Scientists— see Bibliography.)

siles. The nuclear arsenals of the world today exceed the wildest projections of the "doomsayers" of 1950.

Some limitations have been placed on this arms race. They are significant ones and therefore warrant further discussion. We have detailed their provisions in the previous chapter.

The Limited Test Ban Treaty

This treaty, signed by the U.S., Great Britain, and the USSR in 1963, prohibits all nuclear testing in the atmosphere, in space, and under water. Prior to this agreement, the tests conducted in the atmosphere and under water—usually in shallow water—lofted tremendous amounts of radioactive gases and particles into the stratosphere, which then traveled around the world and caused a slow fallout of radioactivity throughout the Northern Hemisphere (see Table 3). Much of the exposure was the result of external radiation, but some strontium-90 got into soils; was absorbed by plants, which were then eaten by cows; and finally showed up in the milk that babies were drinking. Horrified, the two superpowers got together and banned the tests that produced such contamination.

Now, this is an exceedingly interesting phenomenon. Here were two nations, armed to the teeth with fantastically destructive nuclear warheads on ballistic missiles, and they were worried about some radioactivity about equal to background levels. There is an air of unreality, of massive schizophrenia, in a world that reacts almost immediately to a little atmospheric radioactivity but does essentially nothing when faced with the prospect of the death of civilization in a nuclear holocaust.

But wasn't the Limited Test Ban Treaty a crucial first step toward nuclear disarmament? Yes and no. On the negative side, the treaty simply moved all nuclear testing underground. A decade later, the Threshold Test Ban Treaty of 1974 limited such underground tests to 150 kilotons, and one might think that this would have slowed the arms race. But it was still possible to test *models* of larger weapons. The increased numbers and sophistication of weapons in the world's nuclear arsenals demonstrate that neither treaty has been much of an impediment to nuclear-weapon development.

There were, however, some very positive aspects of the Limited Test Ban Treaty. For one thing, it showed that the U.S.

272 □ Chapter 16

and the USSR could arrive at a nuclear arms agreement and abide by it. Thus far, the treaty has not been violated by either side. Also, by banning tests from the oceans and outer space, the treaty limited the range of nuclear-weapon activities, much as was done by the Outer Space Treaty, the Seabed Treaty, the Latin-America Treaty, and the Antarctic Treaty. *The Limited Test Ban Treaty stands as a model of what the superpowers can do if they put their minds to solving a problem that threatens them both.*

The SALT Treaties

The first series of Strategic Arms Limitation Talks (SALT I) was signed by both sides in 1972. It placed limits on the numbers and types of intercontinental ballistic missiles. This treaty was honored by both sides. SALT II, an extension of SALT I, was signed by the Soviets and the Americans in 1976, but it was never ratified by the U.S. Senate. President Reagan called the treaty "fatally flawed" but observed its provisions until November 1986, when we deployed our 131st bomber equipped with cruise missiles.

These treaties have been important in keeping some control over the proliferation of ballistic missiles. They have not stopped the growth of nuclear arsenals, but they have slowed their *rate of growth*. At the same time, however, they have encouraged the development of more sophisticated weapons, the philosophy being that, if each of us can have only so many weapons, then I shall make my weapons better than yours.

The Antiballistic Missile (ABM) Treaty

Signed at the same time as the SALT I treaty, the ABM Treaty prohibits antiballistic missile forces except for one site in each country. The Soviets retained the system they had already set up around Moscow, which now includes radars and 100 defensive missiles—the upper limit permitted. We set up a system around a Minuteman ICBM silo complex at Grand Forks, North Dakota, but dismantled it a year later.

Here again, we have a remarkable situation. Present administrations complain about extensive Soviet ABM activity. Yet the site around Moscow is perfectly legal. Furthermore, it could easily be overwhelmed by a massive attack. We, on the other hand, don't even have the one site we are permitted to have! We could have erected a system around Washington, D.C. Why didn't we? For a very good reason: We decided that ABM forces were not worth the trouble and expense because they could be so easily penetrated. We also felt that *the development of ABM forces represented an escalation of the arms race*, and this is why we opposed the Soviet desire to develop such systems. Now, with our Star Wars plans, we are defending quite the opposite position!

Mutual Assured Destruction (MAD)

Once the superpowers had assembled significant nuclear arsenals, the United States established a fundamental policy of "massive retaliation." It was enunciated by Secretary of State John Foster Dulles during the Eisenhower administration in 1954, as a policy of *launching all strategic nuclear forces on Soviet, Chinese, and East European cities* upon the initiation of nuclear war with the Soviet Union. This was an apocalyptic policy. As the Soviet Union could presumably do the same to us, it soon became known as **Mutual Assured Destruction**, or **MAD**. The idea is that one side would not dare attack the other because that action would ensure its own destruction. This is a very negative situation, and one that leads to paranoia on both sides. Any nation that is attacked is committed to the destruction of the other nation, *even after it may have been largely destroyed itself*. The acronym *MAD* is quite appropriate!

We have never been comfortable with the MAD policy, considered by many to be totally immoral. Later administrations tried to soften the policy. The Kennedy and Johnson administrations adopted the idea of "flexible response," in which we diversified our weapons systems and tried to make them less vulnerable to attack. The emphasis moved toward attacking enemy military forces rather than civilian populations. However, the threat of eventual attacks on cities was still there, should the enemy not surrender after an attack on its military forces. Early in the Reagan years, plans were made for a protracted nuclear war. Electromagnetic-pulse explosions would be followed by attacks on political and military control systems: the capitol city, air-defense command bunkers, and such. If such attacks were successful, the enemy would find it difficult to retaliate. Should it appear that he might retaliate, the next targets would be military bases and missile launchers. Should that action fail to stop the war, cities would be targeted. This is a neat little plan, sometimes referred to as *limited nuclear war.* The problem is that hardly anybody feels that it would result in anything other than all-out nuclear war. One need only consider how a general in Texas might respond, knowing that Washington, D.C., and everyone in it, had been obliterated. Limited nuclear war is felt by many to be an extremely dangerous concept—it holds out a false hope of victory for the side that initiates it.

Yet MAD appears to have worked. So far. This is of course a self-fulfilling prophecy. As long as we don't have a war, one can say that MAD has worked. If we have a war, there will be no one around to say that it didn't work.

One virtue of the MAD philosophy is that it encourages a parity of nuclear arms, both sides being willing to limit the numbers and types of weapons. *A corollary of MAD is that defensive systems cannot be permitted.* A defensive system gives the side that deploys it an advantage that tends to undercut the strategic balance. This is the most fundamental criticism being made of the Star Wars idea. The basic argument for Star Wars is that, being defensive, it is a more humane solution than MAD. But is this really so?

Strategic Defense Initiative (SDI): "Star Wars"

On 23 March 1983, President Reagan went before the American people and called on our scientists "to give us the means of rendering these nuclear weapons impotent and obsolete," by building a "nuclear umbrella" in space to shield the U.S. from incoming missiles.[1] It is a very attractive idea, and in his speech, Reagan raised the moral point that it is better to *save* lives than to *avenge* them. The President's call produced immediate reaction throughout the country, and within a year, powerful protagonists developed. Those favoring the concept called it the **Strategic Defense Initiative (SDI)**, and those opposing called it **Star Wars**.[2] The basic concept is simple, but its implementation is not. It involves a complex command network that directs space-based devices to destroy incoming ballistic missiles. Its political and social implications are also complex. I will discuss this matter further in the next chapter; here I present the technological background and some major political issues surrounding the implementation of SDI.

The Technology

First, let us consider the major technical aspects of the plan. Figure 41 illustrates the events that occur during the flight of a MIRVed (or multiple-warhead) ICBM. The **boost phase** is the period of time when the rockets are firing to launch the missile, speed it up, and set it on its trajectory, the rockets shutting off when the missile is above the atmosphere. The missile then coasts all the way to its target. In a MIRVed missile, however, during the **postboost phase**, the front section of the rocket, called a *bus*, releases the multiple warheads one-by-one, each time changing slightly the direction in which it is pointing—by using small maneuvering rockets—so that each warhead is directed at a different target. While traveling toward the target, *each warhead* is accompanied by a number of **decoy warheads**, containing no explosive, and a large amount of **chaff** (metal

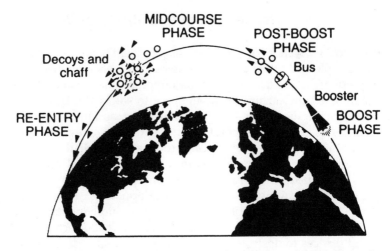

Figure 41. Diagram of the flight of a MIRVed (multiple-warhead) ballistic missile. (From Tirman, 1986—see Bibliography. Courtesy of Union of Concerned Scientists —see Bibliography.)

wires or strips, like ribbons of aluminum foil). The purpose of the decoys and the chaff is to confuse enemy radar and anti-missile defenses. Upon reentry into the atmosphere—the **reentry phase**—at a point only about 50 miles from the ground target, the fragile decoys and chaff burn up from friction with the atmosphere, while the heavy and heat-shielded warhead proceeds to the target. This entire process takes a maximum of 30 minutes; the boost phase lasts 3 to 5 minutes, and the reentry phase 30 to 100 seconds.

The easiest way to shoot down, or disable, a warhead is to hit it in the reentry phase, when it is close at hand, thus facilitating aiming. Furthermore, one is not distracted by the decoys and chaff, which are no longer present. Such a defense can be mounted with ground-based rockets, and does not differ in principle from the antiballistic-missile (ABM) systems that we discussed earlier. A disadvantage is that the response must be very rapid and accurate, for only about a minute is available. Also, one has to shoot down *all* of the multiple warheads released by the original missile.

The next easiest defense is to destroy the missile during the boost phase, thus eliminating all the multiple warheads at once. Satellites can readily detect the missile in the boost phase, because of the tremendous heat and light emitted by the booster rocket.

Missiles can be destroyed by weapons that fall into two classes: (1) **directed-energy weapons**, which emit either a powerful beam of light from a laser, or a beam of atoms or nuclear particles from an electric gun; and (2) **kinetic-energy weapons**, which shoot projectiles. The directed-energy weapons send their beams at or near the speed of light (700 million miles per hour), but the kinetic-energy projectiles are far slower. Among the latter is a device called an **electromagnetic rail gun**, which can shoot as many as 10 tiny projectiles per second at very high velocities reaching thousands of miles per hour; such small objects can be extremely damaging when they are traveling so fast.

These weapons would generally be *space-based*, but some could also be *ground-based*. For example, to attack in the boost phase, a laser beam could be sent from the ground to a large mirror in space, where it would then be reflected to another large mirror over enemy territory, and thence down to the missile being launched. But imagine the technical problems of aiming the mirrors! To attack in the reentry phase, electromagnetic rail guns could shoot projectiles at incoming warheads—or standard ABM rockets could be used.

Technical Problems

Will the apparently noble concept of SDI, as enunciated by President Reagan, in fact work, and will it save us from nuclear war? Many people say no. James C. Fletcher, appointed by President Reagan to head NASA and to chair a panel on Star Wars, concluded that "There is no such thing as a nuclear umbrella."[1]

Five former senior policymakers—Harold Brown, Melvin Laird, James Schlesinger, Brent Scowcroft, and Cyrus Vance—suggested to the Reagan administration in 1986 that it reach agreement with the Soviet Union to put off the testing of space-based weapons for up to 10 years.[3] Why do such people resist SDI? The answer is complex.

Space-based weapons require a fantastically complicated information network. Satellites in orbit must detect missiles and must transmit information about their position, velocity, and trajectory to the "gun" that will shoot the missile down. They must be able to tell the gun exactly where the missile is. In the boost phase, the missile is some distance ahead of the heat and light of the rocket exhaust, and this distance changes with time. The gun must be aimed with extreme accuracy. And all of this must be done for perhaps thousands of missiles within a period of a few minutes.

Finally, the information systems can never be completely tested without a nuclear war. No computer specialists will guarantee anything about their programs if they haven't had a chance to test them and remove the inevitable "bugs." These bugs can be extensive and highly unpredictable in such a complex information network. After a two-year study, the Congressional Office of Technology Assessment concluded that "the first (and presumably only) time the [ballistic missile defense] system were used in a real war, it would suffer catastrophic failure."[4] In short, the technical problems of *command and control* are mind-boggling even for professional military experts, many of whom doubt that the problems can be solved at all.

Another development that could negate an SDI system is that it is possible to design ICBMs that have a *shortened boost phase* of less than a minute. Such weapons shut off their rockets before reaching the top of the atmosphere.[5] Their trajectories therefore cannot be accurately determined from space, because of atmospheric interference and the very short time in which measurements can be made. Furthermore, as most directed-energy beams would be absorbed by—and would therefore have difficulty penetrating—the atmosphere, the much slower kinetic-energy weapons would have to be used.

The space-based system is totally ineffective against cruise missiles. Equally evasive are ballistic missiles launched on a "low trajectory," such as missiles fired by a submarine near an enemy shore.

The problems of command and control, and of the vulnerability of space-based systems, have led many proponents of SDI to admit that, with our present arsenals, which provide tremendous overkill, *enough missiles would get through to devastate the nation*. Therefore, SDI proponents are now concentrating on defending military targets rather than populations, thus giving us the chance for retaliation, which might deter the enemy from launching a first strike. Also, because of the technical difficulties with devices in space, the military is now planning to use mostly ground-based systems that act at the reentry phase. Such ground-based systems, however, involve a major upgrading of ABM defenses, which violates the ABM treaty. The original idea of SDI has thus been vastly diluted.

There are also problems concerning the Soviet response. The complex system of sensing satellites, command satellites, and space-based guns can be easily *shot down* by the enemy, either from the ground or from satellites. As Harvey Lynch, of Stanford University, said at a recent symposium of the American Association for the Advancement of Science, "Compared to destroying a missile, destroying a satellite by SDI is a rather trivial task: satellites move on paths which can be predicted long in advance, and generally speaking they are rather vulnerable targets."[6]

The Soviets would not sit around for a couple of decades while we deploy this threatening system above them. In fact, they have already been accused by current administrations of having an antisatellite system. But it is crude and does not pose a significant threat to Western military or civilian satellites. Furthermore, the Soviets have offered to stop deploying the system in exchange for assurances that the U.S. will not deploy its own.[7]

And, if we build SDI, why would the Soviets not *also build their own SDI*? Although many would argue that we will stay ahead of them because of our superior computer and weapons technology, the existence of even an inferior Soviet SDI would greatly diminish the advantage created by our own. One might think, as President Reagan apparently did, that, if we both have a defensive shield, we are both safer. But in fact, such shields may make it possible for either side to initiate a massive attack that would partially penetrate the enemy SDI, and that at the same time would destroy many of the enemy satellites designed for an SDI response. The enemy would then have to retaliate from a *depleted* and partially blinded arsenal against an *intact* SDI system. Thus, the impetus for a preemptive first strike may actually increase if both sides have an SDI system. As defenses grow, the situation shifts from a deterrence equilibrium to a "mutual preemption equilibrium," in which both sides believe it better to strike first than to deter the other side with threats.[8]

Political and Economic Problems

Sixty-five hundred American scientists, including a majority of the physicists in the nation's top 20 university physics departments, have refused to work on Star Wars projects,[9] because they consider such activity immoral or misguided. What reasoning leads to such a stance?

Any upgrading of ABM defenses, on the ground or in space, runs afoul of the ABM treaty. The primary rationale for the ABM treaty is to limit the escalation of the nuclear arms race. Building better defenses does not necessarily increase security: often it simply offends the enemy, who then devises ways around your defense system. Thus, one becomes involved in an ever-increasing cycle of offense-defense, and the nuclear arms race grows and grows. The President's Commission on Strategic Forces (the Scowcroft Commission) stated, in its final report in April 1983:

> One of the most successful arms control agreements is the Anti-Ballistic Missile Treaty of 1972. . . the strategic implications of ballistic missile defense and the criticality of the ABM Treaty to further arms control agreements dictate extreme caution in proceeding to engineering development in this sensitive area.[10]

Other treaties would be violated by Star Wars deployments. One directed-energy concept involves the explosion of a nuclear weapon in space, the X rays of which could activate a device that would produce a powerful X-ray laser beam. The whole thing would blow up in a microsecond, but the laser beam would be produced and escape before this happened. However, the Limited Test Ban Treaty prohibits nuclear explosions in space, so this system could not be tested without violating that treaty. In addition, the Outer Space Treaty prohibits all weapons of mass destruction from outer space, which would include many of the SDI space-based devices.

It must be recognized that many of the SDI devices could be directed against ground targets in an *offensive* mode. Think of what a laser beam might do to a petroleum-storage complex! Peter D. Zimmerman, of the Carnegie Endowment for International Peace, stated:

> SDI will not produce weapons which only destroy other weapons. They will also serve as strategic arms, almost perfectly suited to strikes against population centers, or as instruments of coercion and destruction.[6]

In addition, as we have already noted, SDI systems may be used to destroy an enemy's reconnaissance and early-warning satellites; this could be done immediately before a first-strike attack. In spite of President Reagan's early assurances that SDI would be purely defensive and that we would even share our knowledge of the technology with the Soviets, there is little doubt that the U.S. Defense Department is aware of these offensive possibilities, having stated:

> We must achieve capabilities to ensure free access to and use of space in peace and war; deny the wartime use of space to adversaries. . . and apply military force *from* space if that becomes necessary.[11]

It is no wonder that the Soviet Union is worried about Star Wars!

The Soviets, incidentally, have conducted no known tests of antisatellite weapons since 1982 and have vowed to perform no further space tests if the United States also refrains from doing so.[12]

A final note on economics: President Reagan's "Fletcher report," produced in late 1983 by James Fletcher, former chief of NASA, estimated a 5-year price tag of $26 billion for research and development, which would stretch to about $70 billion over 10 years. The completed SDI system will involve lofting 100,000 *tons* of equipment into space, requiring some 5000 shuttle flights, and may, according to James Schlesinger, Secretary of Defense under Nixon and Ford, cost close to $1 *trillion*.[13] And this does not even address the cost of maintaining the system once it is in place. Excessive military spending is already weakening the social and economic fabric of our nation. Spending on Star Wars would greatly accelerate this process. Yet, in spite of the relaxation of the Soviet threat because of the recent political upheaval in Eastern Europe and the USSR, President Bush's budget request of January 1990 asked for an *increase* in SDI spending.

In some future world, when long-range nuclear weapons might be limited to perhaps 100 on each side, one might permit the superpowers to have an SDI capability for shooting down a few incoming missiles. Such an SDI capability would not seriously interfere with the deterrent function of a limited missile arsenal, but it would provide protection against either a missile launch by mistake or an attack by a small nation with only a few missiles.

In summary, the arguments against SDI, or Star Wars, are that:

1. It will always let enough nuclear weapons through to destroy us.
2. It is easier and cheaper for the enemy to shoot down such systems than it is for us to build them.
3. Any defense is in the last analysis an offense.
4. Our economy cannot stand the strain.
5. The extension of nuclear-war weapons to space is a dangerous and unnecessary broadening of weapon deployment.

The last point is the most important of all. Whether or not Star Wars can be made to work, and whether or not we can afford to try, it is exceedingly dangerous at this time for us to be *extending* the scope of nuclear armaments instead of *reducing* it.

17 New Perspectives

The nuclear arms race is a frightening spectacle. When one recognizes that the missiles on a single modern nuclear submarine can destroy most of the large cities and military bases of the U.S. or the USSR, one wonders why we continue to accumulate more and more of these increasingly sophisticated and deadly weapons. The answer appears to be *fear* and *irrationality*—the very same factors that inhibit our development of nuclear power.

There has been a "sea change" in Eastern Europe and the USSR since 1989. The Soviet Union is no longer the monolith it once was and undoubtedly is not as great a military threat. But this change has not been reflected in significant reductions in U.S. nuclear arms. One reason for our inaction may be that we cannot yet tell whether the Soviet Union will be a nuclear threat in the future. But another reason we have not changed our stance is simply sheer inertia: we have become used to great nuclear arsenals. We seem to have forgotten how dangerous they are. It is worth spending a little time examining why we built these arsenals in the first place. Then perhaps we can address the question of how to reduce them.

The Evil Empire

Our primary fear has been of the Soviet Union. *There are very good historical reasons for our distrust of the USSR.* The Soviets have always had their own share of paranoia, growing out of the general antagonism of other nations to communism. This antagonism was exemplified in its most extreme form by Hitler's invasion of the Soviet Union. The Russians have also had an inferiority complex, from the recognition that they were not a completely modern nation and had some way to go to achieve the economic and social success of countries like the U.S. and France. Near the end of World War II, President Roosevelt, recognizing this insecurity, tried to alleviate it by making concessions to the Soviet Union, primarily at the Yalta talks. The southern half of Sakhalin Island and the Kurile Islands, former Japanese possessions, were ceded to the USSR. The borders of Poland were shifted westward, and a large eastern portion of Poland was given to the Soviet Union. These changes provided the USSR with additional ports on both the Pacific and the Baltic Sea.

In 1946, only one year after the first nuclear explosion, the U.S. proposed to the United Nations a revolutionary plan—the Baruch plan—to turn over all control of nuclear production to an international body. Had this happened, there would have been no hydrogen bomb and no nuclear arms race. But the Soviets vetoed the plan, as they did almost every initiative taken in the UN in the late 1940s to promote world government.

The territorial concessions of Yalta, which were opposed by Winston Churchill, failed to assuage Stalin's appetite. During World War II, he swallowed Latvia, Lithuania, and Estonia and he occupied Mongolia and North Korea. After the war, he continued to occupy the countries of Eastern Europe, which had been promised free elections at Yalta, and he promoted the invasion of South Korea. In sum, Roosevelt had bent over backward to make the world less threatening to Russia, and Stalin responded by grabbing everything he could and threatening every nation he could. Without the Marshall Plan and the Truman Doctrine, all of Western Europe might have become communist. By the time of Stalin's death in 1953, the U.S. was frightened to death of the Soviet Union, a mindset that provided rich soil for the excesses of McCarthyism. Nor was that the end. In 1961, under Nikita Khrushchev, the Soviet Union built the Berlin Wall. It thus became the only nation in history that has ever built a wall *to keep its own people in*, rather than to keep invaders out. Khrushchev tried to install nuclear weapons in Cuba, sparking a crisis that almost resulted in nuclear war. Since then, the Soviet Union has continually promoted communist revolutions throughout the world—such as in Vietnam, Angola, and Ethiopia—using tactics of subversion and violence.

It is therefore quite reasonable that we should have developed a deep distrust, and even fear, of the Soviet Union. But we, like the Soviets, can also be irrational in our fears. McCarthyism was a prime example of a paranoia that not only accomplished nothing but had a net negative effect: it destroyed the careers of many innocent people and established an irrational anti-Soviet mindset from which we are still trying to recover. The widespread perception in America today that we cannot trust the Russians to keep nuclear treaties, abetted by the Reagan administration, is a further reflection of our paranoia. We documented in Chapter 15 the fact that the Soviets have kept most of their nuclear treaty obligations. The few purported violations that stand up to scrutiny are debatable and not of great importance and may well be balanced by our own violations. Furthermore, we don't really *have* to trust the Russians, since today's methods of verification of treaty compliance are so effective.

Why, in view of their aggressive behavior since World War II, have the Soviets kept these treaties? I believe it is for three reasons. First, it is in their best interest to keep these treaties. If they don't, the nuclear arms race would spiral out of control, and in such an event, the U.S. could probably maintain superiority. Second, the Soviets have had increasingly deep economic problems: they cannot tolerate further massive expenditures for arms. Third, and most important, *the Soviets appear to recognize the futility of nuclear war*. In this respect, they seem more rational than we. A nuclear war between the U.S. and the USSR would simply destroy both nations.

After World War II, when the Soviet Union was at its most bellicose, there were proposals within the American government to wage a "preventive war" against the Soviet Union. Our nuclear monopoly at that time would have made it a short war, with relatively little damage to the USSR. The case of Japan shows that the dropping of just one or two nuclear bombs on a nation that has none may lead the most recalcitrant nation to surrender. Had we conducted such a war, we could have prevented the nuclear arms race (remember that *we* proposed the Baruch plan), as well as the virtual enslavement of Eastern Europe.

But we did not conduct such a war because the concept of deliberately bombing a nation with which we were at peace was abhorrent to the American mind. Instead, we followed the tactic of resisting the Soviets in every possible way—militarily, economically, and politically—*in the hope that they would eventually become more tractable.* This policy was known as the *doctrine of containment.*

This approach finally seems to be paying off. There has been a dramatic change in the attitudes of the "nation of steel"— a remarkable softening of the classical Soviet intransigence and belligerence. Eastern European nations are throwing off the communist yoke. Even the Soviet Union seems prepared to give up its commitment to communism. These astonishing developments deserve nothing but encouragement.

But communism is not yet dead, as was pointed out in a letter to the *New York Times*, on 7 January 1990, by Harriet E. Gross, a professor of sociology at Governors State University, in Illinois:

> What we are seeing is not the vindication of capitalism, but the surrender of authoritarian, corruption-ridden police states—whose governments failed as much from mismanaged as from centrally managed economies. . .
>
> A distinction must be kept between economic systems (capitalism-Communism) and political systems (democracy-totalitarianism).

The important question is whether the Soviet system will cooperate with us merely on an economic and social level, and continue a vigorous military confrontation. We must recognize that *the Soviet Union is the only nation in the world with which we have had nuclear confrontation*. We do not have similar problems with China, for example, which is as communistic and potentially as powerful as the USSR. This difference suggests that what we really face is the threat, not so much of international Communism as of Russian imperialism, which has been around for centuries. Gorbachev has clearly indicated that he recognizes the danger of a nuclear arms race. Yet the Soviet Union has not so far cut its strategic nuclear arsenal in any significant way. Neither, in fact, has the U.S.

Still, we know that both arsenals represent ridiculous overkill. McGeorge Bundy, special assistant to the President for National Security Affairs from 1961 to 1966, stated, in 1990:

> There will never be enough [weapons] to win. . . . What you need is enough to deter, as Dwight Eisenhower was the first president to say. . . . The stalemate of today would be a stalemate even if one side were to go on building while the other cut back by half its many survivable warheads.[1]

Indeed, if each side had a *tenth* of its present nuclear power, it would still be an absolute threat to the other. Our virtually undetectable nuclear submarine force alone constitutes an overwhelming deterrent.

In view of this new world appearing behind the Iron Curtain, as well as the present overkill in nuclear arsenals, *there is now no logical reason why we should not attempt to vastly decrease the threat of nuclear war by making sweeping new treaties with the USSR.* This process began, in fact, in a minor but significant way, with the signing of the INF treaty in December 1987.

Of course, there are those who will say that our vast military machine is what brought the communist states to heel. This is probably not true, although we may never really know for sure. It is much more likely that the Soviet system simply fell apart because of internal strains, mostly economic. This almost certainly would have happened just as readily if we'd had half our present nuclear arsenal.

Intermediate-Range Nuclear Forces (INF) Treaty

One of the most remarkable developments in the Reagan administration was the 1987 signing of the Intermediate-range Nuclear Forces (INF) Treaty. To many, it seemed that Reagan was making a complete about-face, veering all the way from the early days, when he spoke of the "evil empire," to putting his arm around General Secretary Gorbachev in Red Square after having signed *the first treaty that actually bans a whole class of nuclear weapons*. Others feel that this was simply the fruit of Reagan's proposal, introduced as far back as 1980, of the "zero option": the reduction to zero of intermediate-range nuclear weapons in Europe. We may never know what really happened. It is clear, however, that Gorbachev is a man of a new mold, never before seen at the head of the Soviet Union, and that much of the credit for this treaty lies in his apparent conviction that nuclear weapons are useless and a waste of any nation's wealth. It is not so clear how much Ronald Reagan's views changed, but it is very much to his credit that he negotiated and signed this treaty.

What did the treaty accomplish? It eliminated *all* U.S. and Soviet land-based nuclear weapons of intermediate range (600 to 3000 miles) and of "shorter" range (300 to 600 miles) *everywhere in the world*. Thus, Soviet intermediate-range missiles on the Chinese border were destroyed, as well as those in Europe. It did not affect weapons carried by surface ships, submarines, or airplanes. As Table 9 shows, the numbers are impressive. The U.S. destroyed 464 cruise missiles in Britain, Belgium, West Germany, and Italy, and removed 180 Pershing ballistic missiles, for a total of 644 warheads removed. The USSR destroyed missiles containing 1595 warheads, most of them in the western Soviet

Table 9. Missiles Eliminated by the Intermediate-Range
Nuclear Forces (INF) Treaty[a]

Missile	Range (miles)	Location(s)	Number of warheads
United States			
Cruise	1550	UK, Belgium, Germany, Italy	464
Pershing Ia[b]	500	Germany	72
Pershing II	1100	Germany	108
			644
USSR			
SS-4	1200	Russia	112
SS-12	550	Russia, Asia, Europe	120
SS-20[c]	3100	Russia, Asia	1323
SS-23	300	Asia	40
			1595

[a]Data from the U.S. Department of Defense, Natural Resources Defense Council, Arms Control Association, Jane's Weapons Systems, and International Institute for Strategic Studies (see, e.g., *Nucleus*, Fall 1987, Union of Concerned Scientists—see Bibliography).
[b]Missiles to be destroyed; Pershing Ia warheads to remain in U.S. custody, outside Europe. Both Pershings are mobile missiles.
[c]441 mobile missiles, each with three independently targetable warheads.

Union and in Eastern Europe, but some in Asia. That these weapons were in fact destroyed, and not replaced, has been verified by on-site inspection teams—from the other country—with broad authority, including inspections *without warning*. (It should be recognized that the treaty concerned destruction of missiles, not warheads, which the two nations were permitted to divert to use in other weapons.)

A major positive aspect of the INF treaty is that it removed the threat of extremely short-notice attack on Russian cities. Pershing II missiles in Germany could have hit Moscow about 8 minutes after launch. Such short notice is exceedingly danger-ous, as it leads to dependence on electronic systems of "launch-upon-warning"; that is, the Soviet Union would launch missiles as soon as a computer told it that enemy missiles had been launched. If the computer makes a mistake, then we have a nuclear war that no one wanted.

Other reassuring aspects of this treaty are that (1) it involved a reduction of only 4% of the world's nuclear arsenals; (2) it removed more than twice as many Soviet as American warheads; (3) even with the treaty, some 4000 NATO nuclear warheads remain in Europe, in artillery shells, short-range missiles, and air-launched bombs and cruise missiles; and (4) Western Europe still has its submarines and surface ships, which have nuclear weapons that could attack the Soviet Union proper; French weapons are not included in the NATO numbers. So Western Europe is hardly left defenseless!

But the most favorable aspect of this treaty is psychological. The U.S. and the USSR have finally signed a treaty that has some teeth in it. They have finally signed a treaty that, instead of limiting, actually reduces nuclear arms. This treaty gives a measure of greater security to both the NATO countries and the Soviet Union, in that it lessens the chance of nuclear attack on their homelands. *The signing of the INF treaty in 1987 is the first major step that the U.S. and the USSR have taken toward nuclear sanity since the signing of the Limited Test Ban Treaty in 1963.* That was a span of 24 years.

The INF treaty was ratified by the U.S. Senate in May 1988, the first ratification of a nuclear treaty since 1972, 16 years earlier (Table 8). One could hardly accuse the U.S. Congress of rash actions when it comes to nuclear arms control!

Where should we go now in our progress toward the greater limitation of nuclear arms? Two things that could be done very quickly are a comprehensive test ban and/or a nuclear freeze.

Comprehensive Test Ban

We have gone on piling weapon upon weapon, missile upon missile, new levels of destructiveness upon old ones. We have done this helplessly, almost involuntarily: like the victims of some sort of hypnotism, like men in a dream, like lemmings heading for the sea, like the children of Hamlin marching blindly along behind their Pied Piper.

GEORGE KENNAN, Former Ambassador to the USSR
The Nuclear Delusion (Pantheon Books, 1983)

A **comprehensive test ban** is an agreement to ban all further tests of any nuclear devices. Why is such a ban desirable? First, *it is easy to initiate*. There is no need for prolonged discussions about who will be permitted how many missiles, or what the power of these missiles should be. We just agree to stop testing. This is so simple, it could be done over the telephone!

Second, *it is verifiable*. It is much easier to check whether a nation is exploding no weapons at all than it is to check whether the weapons being exploded are below a certain power level. One of the virtues of the INF treaty is its elimination of a whole class of weapons, which is easier to verify. We have already documented in Chapter 15 the fact that we can detect any nuclear explosion of power greater than 1 kiloton. In addition, our ability to monitor Soviet installations is truly amazing. During the SALT II negotiations, for example, both sides agreed to exchange lists of operational strategic missiles. The Soviets listed 1398, but we had counted 1416. It turned out that the extra 18 were at a test site, and the Soviets did not consider them operational. Thus, our estimate agreed exactly with the Soviet count.[2]

Another important point should be made: verifiability does not have to be perfect. If a few missiles, or a few explosions, should escape our notice, this would have no significant effect on the enemy's ability to enhance its nuclear arsenal.

Third, *it will lead to a lack of dependence on nuclear weapons*. This is the whole point of a comprehensive test ban. It would be difficult to develop new types of weapons with any confidence. Additional weapons of existing types could still be made, but these weapons would gradually become obsolete and confidence in their performance after years of storage would decrease. With time, nuclear weapons become less reliable because of internal chemical reactions and mechanical damage from handling, and because of decay of the tritium used in most modern weapons, which has a half-life of 12 years. American military officials have repeatedly stated our need for constant retesting of existing weapons. Thus, a comprehensive test ban would put a brake on the further development of nuclear weapons, and it would produce a gradual loss of confidence in the weapons that already exist.

So why haven't we achieved a comprehensive test ban? We have actually come tantalizingly close! As noted in Chapter 15, during negotiations for the 1963 Limited Test Ban Treaty, we almost achieved a comprehensive ban, but it failed because we wanted 7 on-site inspections per year, and the Soviets would permit only 3. Part of our hesitation was due to information from the Nevada test site that a certain type of seismic wave, called a *Love wave*, was actually being produced by nuclear explosions. It turned out that this was a false interpretation. Lynn R. Sykes, head of Columbia University's earthquake studies group, commented: "In short, if seismologists had done their homework thoroughly by 1963, the nations of the world might have achieved a comprehensive test ban treaty then."[3]

The Limited Test Ban Treaty of 1963, now signed by 118 nations, called on nuclear-weapon nations to work toward a comprehensive test ban. In addition, the 1970 Nuclear Non-Proliferation Treaty, signed by 134 nations, states that the superpowers will strive for "cessation of the nuclear arms race." The international community has repeatedly called for a comprehensive test ban. Nonaligned nations have summoned the necessary one-third support at the United Nations General Assembly to require a January 1991 conference to convert the 1963 treaty into a total ban on nuclear testing.[4] Proponents argue that such a ban would make further development of nuclear weapons difficult for nations like Israel, South Africa, and Pakistan, which have signed the 1963 treaty.

The problem, as we have noted, is fear and irrationality. Part of our irrationality is the conviction that Star Wars—which requires nuclear tests—is a purely defensive measure. As we showed earlier, it is not. But this attitude prevents us from establishing a test ban. Another part of our irrationality is an unwillingness to look at the historical facts that have occurred *since* World War II. We have a strong tendency to regard the Soviet Union in the same mental framework in which we regarded Hitler. We won World War II in a burst of macho glory. But we were lucky then, and times have changed. It is fatal to regard World War III as simply another World War II. Yet, that appears to be exactly what our Department of Defense does.

The Soviet Union declared a unilateral comprehensive nuclear test ban on 6 August 1985, the fortieth anniversary of the bombing of Hiroshima, and urged the Reagan administration to do the same. It refused. On 7 August 1985, Rear Admiral Eugene J. Carroll, Jr., U.S. Navy (Ret.), wrote in the *New York Times:*

> First, the Administration charges that Moscow broke the last test moratorium, in 1961. False. There was no moratorium to break. In December 1959, President Dwight D. Eisenhower ended the 1958 moratorium by formally stating that America considered itself free to resume testing. Moscow was under no legal or ethical restriction to refrain. . . .
>
> The Administration also asserts that the Soviet Union gained a major advantage over the United States by its surprise resumption of tests in September 1961. False. From Sept. 1, 1961, until the end of atmospheric testing on Aug. 5, 1963, the United States outtested the Soviet Union nearly 2 to 1—that is, 137 to 71.
>
> It is contended that the Soviet Union conducted a spurt of testing immediately before declaring the [Aug 1985] moratorium, thus gaining an advantage over the United States. False. According to Energy Department announcements, America has conducted nine tests, the Soviet Union only four, in all of 1985.

One could feel excused for concluding that the Reagan administration deliberately attempted to delude the American people. Surely, it had access to the same facts as the rest of us.

As for the Soviets having gained any advantage from the current moratorium, they held to it *unilaterally* for *19 months*, before finally being forced, by our obstinate refusal to cooperate, to resume testing on 26 February 1987.

Nuclear Freeze

A comprehensive test ban would stop only the *testing* of weapons; it would not stop the further production of weapons that had already been well tested. A **nuclear freeze** would stop all *production, testing, and deployment* of new nuclear weapons; because production would be stopped, replacements could not be made. Thus, the numbers of nuclear weapons would be frozen at the present level. Because testing could not occur, faith in the reliability of existing weapons would slowly wane, and eventually all nuclear weapons would become obsolete.

The idea of a nuclear freeze has received wide support. Roughly a dozen state legislatures and labor unions have supported it, as well as hundreds of city councils and over a hundred national and international organizations, including the General Assembly of the United Nations. The U.S. Senate and the Soviet Union have both proposed it on several occasions.[5] But such a nuclear freeze is strongly resisted by the American military and its congressional supporters.

Would such a treaty between the U.S. and the USSR eventually eliminate nuclear weapons, since other powers also have them? Figure 42 shows that the overwhelming majority of nuclear testing is done by the two superpowers. It would thus remain for the superpowers to convince France, England, and China to forgo production and testing. It probably would not be difficult to convince France and England, which have already developed enough bombs to feel secure in the interim during which the U.S. nuclear stockpile remained reliable. China is an open question. However, once a superpower agreement had been in place for a few years, the tremendous pressure of world opinion would undoubtedly force even China to acquiesce.

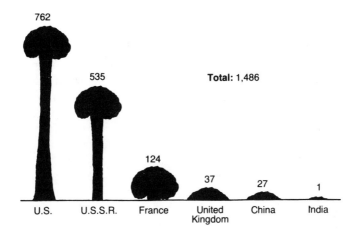

Figure 42. Numbers of nuclear explosions by all six nuclear-weapon-capable countries in the period 1945 to mid-1984. (Sources: U.S. Department of Energy; and Stockholm International Peace Research Institute (SIPRI); chart from *The Defense Monitor* **13**, No. 5, 1984. Courtesy of Center for Defense Information—see Bibliography.)

START and DPB

The treaties described above can be negotiated with relative ease and could be achieved very soon. But they are to some extent stop-gap measures. Significant nuclear arms reductions will require much more detailed agreements. We will discuss two proposals here.

The *Strategic Arms Reduction Talks (START)*, pursued in Geneva since 1981, were envisaged by Reagan and Gorbachev as leading to a 50% reduction in strategic nuclear weapons. Such an agreement would be a giant step toward a reduction in the tremendous overkill that both sides have accumulated in nuclear weapons. It will not come easily because, unlike the INF, it could possibly give one side a distinct advantage. For example, equal cuts in all sectors of the nuclear triads pictured in Figure 32 would leave the U.S. with greatly superior strength in bombers and nuclear subs, and the USSR with greater superiority in land-based ICBMs. However, because of what we know about the attitudes of the bargainers, such an unbalanced agreement seems unlikely. It now appears that a START treaty may be negotiated by 1992. As a part of START, the U.S. has also proposed a ban on all mobile long-range missiles. The virtue of this proposal is debatable. Mobile missiles are hard to target and thus make it difficult for an enemy to carry out a first strike. Nuclear submarines represent mobile missiles, and they tend to stabilize nuclear deterrence.

We have noted that the Mutual-Assured-Destruction (MAD) concept, although it *appears* to have worked for the last four decades, is basically an undesirable strategy, both militarily and ethically. With no defenses, accidental war could break out easily, because there is no way to stop a maverick incoming missile. Furthermore, an attacked nation must retaliate by launching its missiles, even if that nation has already been largely destroyed. Thus, the two superpowers are held at bay by a mutual and debilitating fear.

The Strategic Defense Initiative (SDI) was proposed as a way to ameliorate this state of mutual fear. Defenses against attack *would* provide some security. But SDI would be destabilizing because, in a balanced (MAD) world, the addition of defense on one side provides an advantage, permitting that side not only to suffer less damage in an attack, but also to initiate an attack with less fear of reprisal. However, there are some conditions under which limited strategic defense, such as ground-based ABM systems, would not be destabilizing. Alvin Weinberg and Jack Barkenbus, of the Institute for Energy Analysis, Oak Ridge Associated Universities, proposed a scenario called *Defense-Protected Builddown (DPB)*.[6]

A defense-protected builddown would provide for the *gradual buildup of nuclear missile defenses in parallel with the gradual builddown of offensive weapons*. The result would be fewer nuclear weapons, with less fear of an attack. Carried far enough, it could result in as few as 100 strategic weapons on both sides, an arsenal against which a credible defense could easily be mounted. In such a world, there need be no fear that war will start by accident, or that an attack by a small nation will start a superpower nuclear exchange: one or a few missiles approaching a superpower could be shot down with ease and certainty.

How would one arrive at this ideal solution? There are many possible routes, and a tremendous amount of excruciatingly careful negotiation would have to occur before it could happen. But some first steps, such as the following, could be taken soon and with relative ease. These steps are not exactly those proposed by Weinberg and Barkenbus:

1. Prohibit MIRVed missiles. They are very effective in offense, have little defensive value, and are hard to shoot down.
2. Prohibit weapons with yield above $\frac{1}{2}$ megaton. Without the very large weapons, nuclear winter would be minimal.
3. Outlaw cruise missiles. They cannot be detected by monitoring satellites.
4. Develop terminal ABM systems, that is, ground-based rockets that would shoot down incoming missiles in the reentry phase.

Such steps would, of course, have to be accompanied by thorough on-site inspections. Furthermore, each step should be publicly announced, and it should be agreed that both sides will conform to each step.

But, one may ask, why all this dither and detail? Why not just eliminate all nuclear weapons in one fell swoop? Unfortunately, this will probably never be possible. *The knowledge of how to make nuclear weapons, and to make them quickly, will always be with us.* If the U.S. destroyed all of its weapons, it could be terrorized by another nation that had quickly and secretly produced as few as 10 weapons. But this cannot happen if the U.S. already has 100 weapons and, in addition, the ability to shoot down with certainty a small number of incoming missiles.

Which of these many alternatives should we follow? The

INF treaty, the comprehensive test ban, and the nuclear freeze are all simple, and somewhat simplistic, plans. Their chief virtue is that they get things moving: they begin to reverse the nuclear arms race. We have already undertaken the INF treaty. It would not be unreasonable to follow this with a comprehensive test ban. Such a ban is on nuclear explosions only; it would not inhibit most research on ABM defenses.

A truly long-term plan has to be complex, like the DPB. However, we cannot proceed very far in reducing weapons under *any* long-range plan until we have solved the problems of U.S.–Soviet mistrust.

V   Living with Lions

18 ⚛ Myth III: War Makes Jobs

Defense spending is the most unproductive form of economic investment government can make.

GERALD McENTEE, President
American Federation of State, County and Municipal
Employees (affiliate of the AFL/CIO)

One of the principal impediments to disarmament is the widely accepted notion that arms production is good for the economy— in short, that war makes jobs. Although this is superficially true, in that a new weapons contract will immediately create new jobs, it is false in the long run, producing fewer total jobs, and serving the economy far less well than civilian production.

Weapons and Jobs

Military production actually creates fewer jobs than civilian production. U.S. government figures show that $1 billion spent in nuclear weapons production creates 24,000 jobs, whereas the same money spent on production of civilian goods would create 38,000 jobs. And $1 billion employs twice as many people in civilian service jobs as it does in military production.[1]

Military employment doesn't help national unemployment very much. Many military jobs are for professionals and technicians, where unemployment is relatively low. Of the jobs in civilian manufacturing, 70% require less-skilled workers, but this figure is only 30% for guided-missile production.[2] As weapons become more complex, more money goes into exotic materials and machines, and less into labor.

But the most important point is that *military spending does not increase the civilian standard of living*. If an automobile company builds a truck for a farmer, the truck helps bring food from farm to market, and it may bring the farmer's family to town to buy clothes and to see a movie. These activities raise the standard of living of both the farmer's family and the people who buy the farmer's food. But a military truck is used just to cart soldiers and equipment around and doesn't add anything to the standard of living of the civilian population. A commercial airliner stimulates business dealings and carries people to vacation spots, both of which raise the standard of living of society. A jet fighter does neither. In both cases, the *construction* of the truck or the airplane creates jobs. But the *use* to which the farmer's truck is put creates even further jobs and wealth, and it also adds to the general standard of living.[3]

Finally, military spending is inflationary. It uses up capital, raw materials, and research talent that are badly needed in the civilian economy. As a result, civilian products are more expensive to produce and of lower quality. In turn, prices rise and the standard of living is lowered. Michael Dee Oden, a senior economist at Employment Research Associates, Lansing, Michigan, wrote:

> Mounting evidence, supplemented by several recent findings, seems to indicate that military spending contributes to inflationary pressures, entails employment costs, is associated with low levels of investment, and is a relatively inefficient way to stimulate technical change.[3]

Let us look more closely at this picture. We can compete successfully with competitors anywhere in the world if we make a better and/or cheaper product. How does one do this with high wages that are climbing ever higher in a healthy economy? It is done by *improving the techniques of production*, which is the approach that Henry Ford used when he instituted the assembly line. By increasing the efficiency of production, we increase worker output. Thus, we lower the cost of the product, keeping it competitive. But this increased efficiency does not happen all by itself. It requires intensive effort on the part of large numbers of scientists and engineers.

Where are our scientists and engineers? At least 30% of them are working, directly or indirectly, for the military.[4] These people are putting their tremendous abilities to work designing things like bombers, missiles, and antimissile devices, which are only infrequently used, even in training, and which become obsolete in a decade or two and are thrown on the scrap pile. The efforts of these technicians add nothing to the standard of living of Americans. In fact, it is largely because of this activity that our standard of living is currently dropping.

Before World War II, technologists worked for the military only in wartime. After that war, we demobilized our armed forces and started moving into a normal peacetime configuration. But the failure of the United Nations to bring about the stable world we thought we had fought for, and the uncompromising belligerence of the Soviet Union, quickly led us to make an about-face. We started again to build up our military forces, and we began to increase spending on military hardware. Except for declines after the Korean and Vietnamese wars, this buildup has continued steadily. As it has grown, the civilian economy has suffered. Fewer technically trained people are available to design automobiles and computers. Even the federal government is spending relatively less money on the civilian sector. In 1980, the total U.S. budget for research and development (R & D) was divided equally between military and civilian, at about $15 billion each. In 1989, the civilian budget had grown to about $21 billion, but the military had almost tripled to $41 billion.[5]

Weapons and the Economy

Excessive military spending now produces some of the same consequences as military defeat; that is, it gives foreign governments greater control over the life of the country.

RICHARD BARNET
Real Security, Restoring American Power in a Dangerous Decade (Simon & Schuster, 1981)

In 1989, 30 million Americans lived in poverty—8 million more than in 1979. Among these, 33% of black adults, and 46% of blacks under the age of 18, lived in poverty.[6] At the end of 1987, it was estimated that 20 million Americans were suffering from malnutrition. The gap between the rich and the poor has been steadily widening.[7]

In the 1980s, even after adjusting for inflation, U.S. military spending grew 25%, while domestic programs were cut 19%, and low-income programs were cut 55%.[6] Education and social services are staggering under the cuts. American education is falling behind that of Europe and Japan. Enrollments of students in science and engineering have dropped so precipitously that we are considered to be in a crisis. This "student gap" will severely impact our future ability to compete commercially with other nations.

In this day of concern about research on AIDS and cancer, the American Federation for Clinical Research finds that "In 18 months the Department of Defense spends more on research than the National Institutes of Health has spent in its entire 100-year existence."[8]

In 1985, for the first time in 70 years, the US became once again a net international debtor, and we are now the world's largest international debtor nation, while Japan is the world's largest creditor nation.

This whole picture is a familiar one in modern history. Paul Kennedy, of Yale University, has traced how all the great powers since the fifteenth century have eventually gone into decline because of excessive military spending.[9] The U.S. is headed in the same direction.

It doesn't have to be this way. America has enough wealth and intelligence to provide for those less able to cope in our great competitive society. But we are siphoning away our great intellectual and financial resources on an arms race that is largely a result of paranoia, rather than of reasoned thinking about a sane military defense posture. In our frenetic attempts to "stay ahead" militarily, we have lost control of our economy, chalking up an outrageous national debt and producing great budget deficits every year. And we are now sacrificing education and social welfare. We are not doing nearly enough to educate our young people, to provide vocational training, and to provide jobs. Nor are we caring for the elderly and for the physically or mentally ill. We need to spend less money on arms and more on people. There are many ways in which we can do this.

Figure 43 shows the correlation between the military's share of research and development (R & D) and the rate of manufacturing productivity growth. Our most successful competitors in world markets are spending the least on their military, a fact that supports the idea that high military expenditures drain the civilian economy. We are providing roughly half of the defense of Europe and Japan. Many people would argue that, if these countries spent a larger fraction of their national resources on their military defense, the U.S. would be able to spend relatively less. Reducing our support of the defense of Europe and Japan would help redress the balance, and the American economy would improve.

We need to reduce our military spending. Not only are our military arsenals much larger than they need to be, but this spending is also crippling our economy. We could probably cut our military spending in half without sacrificing our national security. How could we do this?

We have noted several times in this book that the nearly 200 warheads carried by the 24 missiles of a single modern nuclear submarine could destroy most of the large cities and military

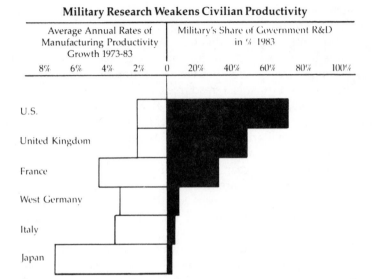

Figure 43. The average annual rates of manufacturing productivity growth in the decade 1973–1983 compared to the military's share of research and development (R & D) expenditures in 1983, for six nations. (Data from the U.S. Office for Economic Cooperation and Development (OECD); chart prepared by the Center for Defense Information—see Bibliography.)

targets of either the U.S. or the Soviet Union. Yet we have over 30 of these subs. Wouldn't 12 be enough, with 6 on patrol at all times? Britain has 4 nuclear subs; France has 6, and the mega-tonnage of France is twice that of Britain. Either of these two nations could devastate any country in the world, with no help at all from us.

We have recently reactivated 4 battleships. These be-hemoths, each representing an operation equivalent to that of a small city in terms of food eaten, power expended, and so forth, became outdated in World War II, and are almost useless today except for "show." When they bombarded the hills behind Beirut, with shells that weighed as much as a Volkswagen, they did little military damage; a single jet fighter would have been more effective because it could have seen what it was hitting. Until recently, we had 12 large aircraft carriers. These ships, tremendously effective in World War II, are now vulnerable to missiles and attack submarines and are useful chiefly against nations with inferior air or naval forces, like Iraq or Libya. The Soviet Union has just built its *first* large attack carrier. Yet, we decided a few years ago to *add another three* to our existing fleet of 12! Such military steps defy reason.

Considering the tremendous overkill of our present nuclear arsenals, we could cut our arsenal in half, and still no nation in the world—including the Soviet Union—would dare attack us. On the other hand, if we and the Soviets agreed to parallel reductions, both sides could cut their arsenals to 10% of the present size and would still be able utterly to decimate each other. In conventional weapons, we need small mobile forces that can be moved quickly to hotspots around the world, and not a ponderous fleet of battleships and aircraft carriers. A fleet with no battleships and 4 carriers, instead of 15, might make sense.

Much money could be saved—some estimate as much as half our military budget—by (1) simplifying the procedures for the procurement of military items; (2) using simpler, standardized weapons that are cheap and reliable, which a soldier could readily be taught to operate; and (3) reducing cost overruns by holding a contractor to the stated bid, instead of paying the costs—whatever they may be—plus a profit. On this last point, Admiral James D. Watkins, Chief of Naval Operations, in Congressional testimony in 1985, stated:

> Today our big contractors average over four times as much profit as a percentage of assets on their defense contracts as on their commercial contracts. Why is that? What is it in the defense business that would warrant four times the percentage of profit? There is nothing.[10]

Considering our vast nuclear arsenals on ships and submarines, and in land-based missiles and artillery, we could probably dispense altogether with nuclear-weapon-carrying aircraft. We have no need for Stealth bombers, whose ability to evade enemy radar is doubtful anyway. We could also lower military personnel costs. For example, we might reduce the current lifetime retirements at half-salary, after only 20 years of service, for personnel who have not served in a combat zone.

In short, there are many ways in which we could cut our military budget without significantly decreasing our ability to defend this nation. Such drastic action would immediately turn our economy around and permit us to restore some of the much-needed aid to education that is essential if we are to maintain a competitive edge in the world.

There is no doubt, as many would argue, that sudden decreases in military spending would throw many people out of work. To prevent this, industry needs to plan for such an eventuality by what is called *economic conversion*. Lest this sound impractical, remember that many firms converted most successfully after World War II. General Motors manufactured most of our mobile weapons during that war, and it converted with relative ease to the civilian production of cars and trucks. The Grumman airplane company started making aluminum buses, which were successful because they had the lightness and strength that Grumman had learned to put into aircraft structures. It has been proposed that companies be required to do economic conversion planning as part of any defense contract. The Soviet Union also supports the idea of planned economic conversion.[11]

We have faced many military challenges since World War II. Clearly, we must keep ourselves strong. However, *strength lies not only in the military power of a nation, but in its economic health, and in the high level of education and the high standard of living of its people*. The best way we can prevail over nations that oppose us militarily is by making America the best place to live in the world.

19 Facts and Fallacies

We have discussed three outstanding "myths" that cloud the thinking of many people on nuclear issues. Why do we have these myths? Often, it is because the communications media give us inaccurate assessments of dangers. But there's more to it than that. We mentioned in the Introduction that many people *mistrust the experts*. This mistrust often results from a misunderstanding of what the experts are saying. Scientists and non-scientists may interpret the same information in different ways. These differences can lead to confusion and distrust.

Fact and Theory

We all have a pretty good idea of what is meant by ordinary physical "facts," and most people are aware of how scientists arrive at "theories" based on those facts. But even in simple cases, the theories that arise may be quite surprising.

Consider the theory that the Earth is round, that is, spherical. This idea actually contradicts common sense. Anyone looking out across a midwestern corn field can easily tell that the Earth is flat. Football fields are flat, and the ocean is very flat. Why, then, do we think that the Earth is round?

First, there is the fact that one can sail around the Earth. Second, if you make a map of a large area, such as North America, the distances between places—which can be measured accurately—do not work out unless you assume that you have a spherical surface. Third, during an eclipse of the Moon, we see a circular shadow, presumably cast by the Earth. Presented with such a *variety of independent facts*, we arrive at the theory that the Earth is round.

No one piece of evidence, however, is sufficient to give us a theory we can trust. The fact that you can sail around the world does not mean it is a sphere: it could be a cube, with rounded edges. And if the Earth were round and flat, like a plate, it would cast a nice circular shadow on the Moon during an eclipse. It is only the wide variety of evidence that makes us believe that the Earth is round.

As time goes on, we devise various tests of the theory, which we can call *predictions*, and they work out correctly only if we assume that the Earth is indeed round. For example, we predict that, when it is daytime in the United States it must be nighttime in China. With radio communication, we can confirm that this is indeed so.

Finally, we have *general agreement* that the Earth is round. Most people are convinced by the evidence; few people with a modern education still think that the Earth is flat.

Now, this is an interesting situation. For we have decided that something is "true" that is quite contradictory to our "common sense." Why are we willing to suspend our natural instincts, which tell us that the Earth is flat? We do it for the reasons outlined above, namely, (1) a variety of independent evidence, (2) predictive power of the theory, and (3) a consensus of opinion that the theory is correct.

There is an important distinction between facts and theory. *Facts don't change*, which is one reason why it takes so long to establish them. Regardless of one's theory about the shape of the Earth, it is a fact—and will always remain a fact—that one can sail around the world. *Theories do change*, although they don't change as readily or as much as most people think. Much of what has been presented so far in this book is not a matter of opinion, because it is based largely on facts. No position that I or anyone else may take on nuclear waste disposal will ever alter the fact that the half-life of radioactive iodine-131 is 8 days. However, the theory that carefully isolated nuclear wastes will not get into drinking water will change as we accumulate more data (facts) on the behavior and permanence of geological formations. However, it will not change a great deal because we already have a lot of these facts. We usually don't propose a theory until there are enough facts to allow the formulation of a reasonable one.

Projection into the Future

There are different kinds of facts. Many are of the type we could call *evident facts*, such as the fact that a child's hand has five fingers, or that water flows downhill. These are facts that you can show or demonstrate to someone. Others are *derived facts*, which at one time were theories, but which are now accepted as fact. When a theory, such as the idea that the Earth is round, becomes very well proven then *the theory itself becomes accepted as a fact*. But it is a "derived fact": you can't easily demonstrate it. You can only infer it from a variety of evidence.

Nonscientists tend to consider derived facts less reliable than evident facts. This is not true. We are even more certain that the Earth is round (derived fact) than we are that children's hands have five fingers (evident fact). Some children have six fingers, a rare but real developmental abnormality. *The circumstance that a fact is derived does not make it less certain.* What makes it reliable as a fact are the criteria we noted above: variety of evidence, predictability, and consensus of opinion.

Most of the facts of modern science are derived facts. Thus, we are all convinced that atoms exist, but no one has ever seen one, and indeed no one *can* ever see one, simply because the wavelengths of light with which we see cannot interact with an atom and be reflected by it. We can, however, get *pictures* of atoms with an electron microscope. The existence of atoms is a derived fact. The roundness of the Earth is a derived fact. The Earth's circling of the Sun is a derived fact. That biological evolution occurred is a derived fact.

In this last case, some people confuse theories about *how* evolution occurred (on which there is much lively debate among scientists) with the fact that evolution *did occur*—on which there is no debate among scientists. The criteria of variety of evidence (which is overwhelming in the case of the occurrence of evolution), predictability (for example, the prediction that human remains will never be found with dinosaur remains in a well-preserved rock formation), and consensus (the world's scientists accept it) make the fact of the occurrence of biological evolution one of our more reliable facts, in a class with the existence of atoms.

One can not only construct a picture of the past, as we do with evolution, but also form a picture of the future. We do this whenever we "plan" something. Thus, an architect's drawing is a picture of the future. At the time when the drawing is completed, the building has not yet been built. Such plans can be quite tentative and therefore not very reliable. But some plans can be so reliable that we can consider them fact. Consider the plans for an airplane. One could say that "this airplane, *if built*, would certainly fly." Now, that would be a fact. The airplane may or may not be built, but if it is built, it will fly.

Two of the major future projections dealt with in this book concern nuclear reactor accidents and nuclear winter. In evaluating the likelihood of these things, we use our three criteria. How well do these scenarios hold up?

For *nuclear reactor accidents*, we have few data, but we do have many years of experience with nuclear reactors. With the exception of Chernobyl, there have been few accidents, with relatively little harm to people. We therefore have a *variety of evidence* that nuclear power is safe. But we also knew from the start what could go wrong and *predicted* that a devastating accident could occur. That the Chernobyl accident occurred was therefore not a surprise to the experts. They *were* surprised at the magnitude of the damage. But the damage was so great because *Chernobyl was not really an accident*: it was an experiment that shouldn't have been done and that went awry. (Some people have compared the Chernobyl "accident" with an airline pilot testing his engines by turning them all off in flight.) So nuclear power, assuming that people don't do dangerous experiments with it, is turning out to be about as safe as the experts have predicted. With a history of so few accidents, it is difficult to project with very great confidence what the likelihood of a dangerous reactor accident may be. But we do have the fact that, Chernobyl excepted, no one has been killed anywhere in the world in a civilian nuclear reactor accident. Civilian reactors have been operating for more than 35 years, so we can therefore say with some confidence that *devastating civilian nuclear reactor accidents (like Chernobyl) are very unlikely to happen*. There is a broad consensus among nuclear engineers that this is so.

Now, it is of the utmost importance that one read this statement exactly as it is. Did we say that a Chernobyl-like accident could not happen? No. We said it is *very unlikely*. What does that mean? It means that we wouldn't expect more than one in something like the next 50 years. Did we say that nuclear reactor accidents are unlikely? No. We said *devastating* accidents are unlikely. Three Mile Island was a very damaging accident, in terms of the cost to the operators, but it didn't hurt anyone, and therefore, from a human perspective, it most certainly was not devastating. Furthermore, note that the statement is limited to *civilian* reactors. Military reactors are often less safe; however, there are many fewer of them, and their numbers are decreasing all the time, so the likelihood of a military accident is also very low. To date, only one military reactor accident (Windscale) has had a significant environmental effect.

But if we hedge our bets like this, aren't we really saying that we're a little confused, and that we don't really know much about reactor accidents? Again, the answer is a resounding *no!* The fact that, for 30 years before Chernobyl, civilian nuclear power reactors had not claimed a single human life is a safety record unparalleled by any other power technology. Nuclear power reactors are, in fact, very safe. But, once again, let me remind you to read the statement carefully. For example, it says nothing about nuclear waste disposal. That is another question.

How about nuclear winter? Here the situation is different, in that we have no actual data from a nuclear war. On the other hand, we have a lot of data from volcanic eruptions, including that of Mount St. Helens. It was observations of Mars that first led Carl Sagan and his co-workers to the recognition that local dust storms could sweep across a whole planet. And we know that great forest fires, such as the one in 1989 at Yellowstone National Park, and firestorms, such as those produced in Hamburg and Dresden in World War II, loft tremendous amounts of smoke to high altitudes. So we do have a considerable *variety of facts*. Computer scenarios can be constructed from these data, using our fairly sophisticated knowledge of Earth's weather patterns, and enabling us to make quite good *predictions* of the probability of nuclear winter. There is still some disagreement about these scenarios (or models), so we still lack a full *consensus of opinion*. But there is general agreement that *nuclear winter will be severe enough to greatly magnify the immediate consequences of a nuclear war*. This statement is not as strong as the one we can make about nuclear reactor accidents, because we lack complete agreement on the severity of the phenomenon.

Probability

A great deal of scientific knowledge is in the form of statements of **probability**, to which we attach numbers. If something is certain to happen, the probability is exactly 1. If there is a 50–50 chance of its happening, the probability is 0.5, and if the probability is 0.1, we mean that there is 1 chance in 10 of its happening.

Now, these statements of probability pertain to what will

happen *on the average*. Suppose one said that the probability that a man of 70 will die in the next year is 0.1. This would mean that 1 out of 10 men arriving at that age would die before his next birthday. But of course some men live to be 100. Have they escaped the rule? No. They are part of the numbers that go into figuring out the average. Some men die long before they are 70, some long after they are 70. The statement is simply that, *on the average*, 10% of those reaching 70 will die before the age of 71. Therefore, if the probability of a nuclear reactor accident is very low, it does not mean that it could not happen tomorrow. It could. But the likelihood is very low.

Probabilities have to taken seriously. Most scientists do take probabilities seriously. Thus, most "literal" people, like scientists, do not worry about flying because the probability of dying in an airline accident is only about $1/30$ of the probability of dying in a car accident—in a trip of the same distance. But some people have great difficulty dealing with such a probability. They concentrate on the horror of the plane crash (forgetting the horror of the car crash) and lots of other factors, such as their being in control of the car, but not in control of the airplane. But these are subjective factors. *They don't change the probability.* If you want to live, fly. If you care less about living, drive. Those are the facts.

Another consideration is that the probability of the occurrence of two independent events is the *product* of the individual probabilities. Thus, if there is 1 chance in 100 of my car brakes failing on a trip, and if there is 1 chance in 10 of my being unable to stay on the road *if* my brakes fail, then there is only 1 chance in 1000 that I will leave the road because of brake failure. In other words, don't worry about this one!

Studies have been made of possible nuclear power plant failures. Suppose it is estimated that the probability of, say, a cooling pipe breaking is 1 in 100. It may be further estimated that, in the event of a cooling-pipe breakage, there is 1 chance in 100 that the emergency backup pumps will not start. This means that there is 1 chance in 10,000 (100 times 100) that a cooling pipe breakage would lead to overheating of the reactor core. In short, it would be a very low probability, and nothing to worry about. However, nuclear power engineers have encountered an interesting reaction from the public: when presented with all of the "backup" systems in a plant, people often say that they "didn't realize that so many things could go wrong"! This reaction represents a complete failure to comprehend that the more backup you have, the *safer* you are. Modern airliners are full of backup systems. As I have said, probabilities have to be taken literally.

Suppose the probability of a nuclear accident is 1 in 1 million. Should we stop worrying? That depends on the *consequence* of the accident. We may attach a number to the consequence, just as we did to the probability. A consequence of 1 would imply destruction of the world. All other events are less consequential and have a consequence less than 1. One may then say

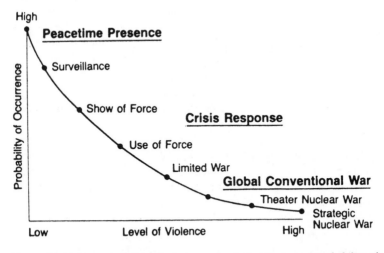

Figure 44. Reciprocal relationship between level of violence and probability of occurrence, as viewed by the U.S. Navy. (*FAS Public Interest Report* **40**, No. 6, June 1987. Courtesy of the Federation of American Scientists, 307 Massachusetts Avenue NE, Washington, DC 20002.)

that the *hazard* of a situation is the product of the *probability* of its occurring and the *consequence* of its occurring:

Hazard = (probability of occurrence) × (consequence)

Thus, the hazard of a nuclear reactor meltdown is not very great because (1) the probability that it will happen is very low, and (2) the consequence is rather low, as probably only one in five meltdowns will release significant radioactivity. A strategic nuclear war, on the other hand, is terribly hazardous (Figure 44). The probability of its occurring is very low (the dot to the far right in the figure), but the level of violence—that is, the hazard—is very high. The product of these two numbers can still be a large number.

Controls

Another aspect of scientific work is the idea of the *control*. Suppose we feed saccharine to 1 million people and 10 of them die. Does this mean that saccharine is dangerous? Maybe not. We need to compare our results with a *control group* of people who are fed ordinary sugar. If 10 out 1 million of them die, then obviously saccharine is no more dangerous than ordinary sugar. *A statement about the hazard of a thing or an event is meaningless without control data.* Thus, we estimate that, on the average, about 150 people will die every year in the U.S. as a result of nuclear power production of electricity (Table 4). That sounds pretty awful. Until, of course, you compare it with a control group that uses coal-fired power, where it is found that 3100 people will die every year. This comparison leaves the choice pretty clear. But it was not clear without the control data.

Another example is the question of leukemia among people living close to a nuclear reactor. A study in England showed that people living near reactors built before 1955 (thus mostly military) had a significantly higher risk of death from leukemia than a control group of people. Thus, one might conclude that old reactors *cause leukemia*. But these studies also showed that people living near more modern reactors have a lower risk of death from leukemia than in a control group. This statistic might lead one to conclude that newer nuclear reactors *prevent leukemia*—an illogical conclusion, and one that would hardly be embraced by the press. So do we shut down the old reactors? Not necessarily. The analysts found that, for the old reactors, the control group showed an unusually low number of leukemia deaths, which suggests that these controls were not well chosen. But there is more to the story—it turned out that the group living near the old reactors, who showed higher risk of death from leukemia, also showed *lower* mortality from other cancers.[1] Thus, it is clear that data without controls are useless. But even data with controls must be interpreted cautiously!

It is important for the layperson to make a sincere attempt to understand exactly what the experts are saying. This is often not easy, but the process is greatly helped by having an attitude of objectivity. In short, if you are to evaluate the safety of nuclear power in a rational way, you must not start out with the conviction that it is unacceptably dangerous. Furthermore, you must be willing to accept the fact that all human activity carries some danger, even staying at home in bed. The important thing is to evaluate the danger relative to the alternatives. In such an evaluation, you must consider the facts. You must consider the probabilities. You must consider the risk versus the benefit. And you must ask yourself if reasonable control studies have been done. A report of higher-than-average leukemia near a nuclear power plant may not mean that the power plant is causing the leukemia. And how many studies have you heard of in which the incidence of leukemia was evaluated near a coal-fired power plant? Such studies are rarely done because no one is excited about the dangers of coal-fired power.

Finally, one must avoid the syndrome of the ostrich burying its head in the sand. People don't like to think about unpleasant things. This is one reason why we have come so close to nuclear war. If people really thought about nuclear war, they would be so aghast at the magnitude of the dangers that they would insist that their congressional representatives immediately and vigorously press for nuclear treaties leading to drastic reductions in these apocalyptic weapons.

20 ⚛ Technology, War, and People

Science is a name we give to the search for fundamental knowledge, as well as to the knowledge itself. As we now know it, science is a relatively modern phenomenon. It blossomed rather suddenly in the Renaissance with the achievements of such men as Sir Isaac Newton and René Descartes. This development has steadily accelerated up to present times. Today, the growth is so phenomenal that roughly 90% of all the scientists that have ever lived are alive today. The advance of science has given us a vast amount of material knowledge.

All advances in material knowledge eventually transform society. *Technology* is the name we give to the application of scientific knowledge. Within historical times, technology first appeared in metallurgy: in the use of bronze to make utensils and the use of iron to make swords and plowshares. It is interesting that, even at these early times, such advances were most clearly manifested in the standard of living and in warfare. The blessings of science and technology have always been double-edged.

Science and technology are often confused. For example, people often say that it was a great scientific achievement when NASA put man into space. But this is not an accurate perception. The scientific achievement occurred 300 years ago, when Isaac Newton developed his theory of gravitation and predicted that any object given a high enough initial velocity could be put into orbit around the Earth. What NASA accomplished is technology. According to Newton's theory, if an object is *not* given a sufficiently high initial velocity, it will follow an elliptical trajectory to another point on Earth. He thus predicted the behavior of intercontinental ballistic missiles. Similarly, Albert Einstein discovered the scientific principle that showed us that the nuclei of atoms have immense energy. But the development of nuclear bombs was the technology, not the science, of releasing this energy.

There has always been the dilemma of the good of science and technology versus their evil. The high standard of living we enjoy today is largely a result of the Industrial Revolution, which in this century alone has killed 100,000 American men just in the mining of coal. Today, with the release of the energy of the atomic nucleus, we face, in an even more exaggerated form, the same problem of the good and evil of science and technology.

Should we put limits on science and technology? It is highly improbable that we would ever want to limit the progress of science. Such limits on free inquiry into the fundamental nature of the world could be as dangerous as Hitler's burning of books in Nazi Germany. It simply isn't consistent with human freedom.

However, it is reasonable to contemplate limiting technology. Indeed, we have already done so: we decided not to build an American supersonic transport, like the Anglo-French Concorde, partly because of the damage to the ozone layer that it might produce. We have decided to limit the kinds of experiments that can be done in genetic engineering; and we have decided not to have any nuclear weapons in space.

In fact, we should currently be putting many limitations on our technology. We are polluting the Earth with the chemical byproducts of our industries. With a little extra expenditure of money, we could convert those wastes into harmless, and in many cases useful, products. We need to limit the amount of sheer *power* that we use in this country. We started doing this under the Carter administration, when Americans significantly lowered their use of heating fuel, gasoline, and electricity. This conservation, plus some economic slowdown, has caused the growth in electric power demand to be much less than was projected some 15 years ago. This development made it possible to stop the growth of our nuclear power industry without adverse effects on the economy. But we still suffer from the environmental consequences of fossil-fuel power, and we should be steadily converting from coal power to nuclear power. In short, limitations on technology, especially on the kind that uses up natural resources and pollutes the environment—mostly the chemical and fossil-fuel industries—is both appropriate and good for the nation. A similar argument can be applied to the development of new generations of nuclear weapons, especially such tremendously expensive weapon systems as the Stealth bomber and Star Wars.

A Moral Imperative

> No man is an Iland, intire of itselfe; every man is a peece of the Continent . . . any mans death diminishes me, because I am involved in Mankinde; And therefore never send to know for whom the bell tolls; It tolls for thee.
>
> JOHN DONNE (1573–1631)

The stakes have now become very high, so high that we find ourselves discussing not the future of humankind, but whether there will even be a future. This is a terrifying prospect, for humankind can eliminate itself only once; there will be no second chance for survival. Most thinking people therefore now believe that *there is no problem facing humanity so important as nuclear war.*

There are many ways to react to such a threat. One is to stick your head in the sand—like an ostrich—and this is exactly what so many of us do. How many people spend as little as five minutes a day thinking about nuclear war? And yet whatever we *do* think about will be of no importance whatsoever if we have a nuclear war. As Dr. Helen Caldicott, of Physicians for Social Responsibility, said:

> By ignoring the true reality of the imminence of nuclear war, we are practicing passive suicide. Our society is in the grip of a pervasive mental illness that will lead to its death.[1]

One who chooses not to be an ostrich may decide to go to the other extreme—something we humans seem extremely prone to—and oppose all armaments, rail against our military, and argue for peace at any price. Is this the answer?

I think not. Although we must recognize the horror and the total unacceptability of nuclear war, we can lessen the danger of nuclear war only through a full awareness of how we got where we are today. In short, we must take a *sane* approach to this, the most important problem that humans have ever faced.

We described earlier some of the history of our nuclear weapons technology. The evidence shows that there are very good reasons why the U.S. (1) *developed* nuclear explosives: the specter of Hitler with nuclear bombs was one that no one wished to contemplate. There are also good reasons why the bomb was (2) *used*. The atomic bombings ended World War II immediately. They killed 175,000 people[2] but probably saved millions. The deaths of these people were horrible, but no more so than the deaths of the far greater number of victims of the fire bombings of Tokyo and other Japanese cities. Finally, there are good historical reasons, which we outlined earlier, why our nuclear forces have been (3) *deployed* against the Soviet Union.

But should we continue to deploy these deadly forces against the Soviet Union? Can we reverse the trend to larger and larger nuclear arsenals, more and more sophisticated weapons, and more chance of nuclear war by accident? In view of the new liberal face of the Soviet Union, can we now relax our guard? The answer lies in a new assessment of our whole military posture. We must take a fresh look at the Soviets and our relations with them.

First of all, the Soviets have demonstrated amply that *they will abide by nuclear treaties*. We have often accused them of violations, but these violations have been few and debatable, and we ignore our own violations. In addition, it was the Soviets, not we, who took the unprecedented step of *unilaterally* stopping all nuclear testing for 19 months, during which time we conducted 22 underground tests. There is nothing that puts a nation's development of nuclear arms in greater jeopardy than cessation of testing. Yet, the Soviets did it. What more could we ask?

Second, *the Soviets fear any kind of war*. They have always felt that they were surrounded by enemies. Although some of this attitude may be dismissed as paranoia, much of it derives from their history, during which they have been constantly beset by invaders. In both world wars, their losses of civilians and military personnel were vast, and bear no comparison to our own losses (Figure 45), which included virtually no civilians. There are monuments all over the Soviet Union to this devastation of their population. They remember.

Furthermore, although the Soviets have undoubtedly been aggressive everywhere in the Third World, they have not actually seized any territory since the postwar days of Eastern European occupation. They came close to it in Afghanistan but have now retreated from that adventure. Their influence in the Third World is much less than it was 20 years ago, countries like Egypt now being clearly in the Western sphere. And the specter of international communism's engulfing the world now seems quite unlikely, in view of the newfound freedom of Eastern Europe and the restructuring of the USSR. The Soviet Union is clearly no longer the military threat that it was just a few years ago.

Third, *the Soviets are human beings*. This may seem a trivial statement. But it is very important.

On a chilly day in January 1983, an attractive middle-aged American nurse walked into the Soviet consulate in San Francisco. This was not a common occurrence at the height of the Cold War. What subversive thing was she up to? Sharon Ten-

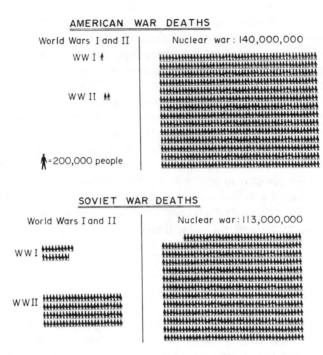

Figure 45. American and Soviet war deaths. (Adapted from charts by Arthur Kanegis, Center for Defense Information—see Bibliography. Nuclear war estimates by the Military Strategy and Force Posture Review, Presidential Review Memorandum #10. From Chivian, Chivian, Lifton, and Mack, 1982, p. 11—see Bibliography.)

nison merely wanted to know if she could get a visa to travel in the Soviet Union *in order to talk to the Soviet people face-to-face about the future.* She had no agenda. She represented no constituency. She did not want to talk politics with Soviet citizens, just to meet them, get to know them, and show them that Americans are really not such bad people. In return, she simply wanted to try to learn what they were really like. The Soviet officials were very surprised by this request. But they told her that she could in-

deed travel in the USSR and go wherever she wished, provided
that she avoided secret military areas. Now it was her turn to be
surprised, for she next went to the FBI and was told never to go
anywhere in the USSR without supervision, and that the Soviets
would not permit her to talk to private citizens without prior
permission. She later learned that U.S. agents across the street
from the consulate had photographed her entrance into the
building and had documented her visit to the vice consul.

The short of a long story is that she went to the USSR, was
amazed at what she found, and has returned dozens of times
since. She has cofounded a group called Center for US-USSR
Initiatives.[3] It is staffed by volunteers and represents an extended community of over 12,000 people in San Francisco, with
sister organizations in Chicago, Tucson, Los Angeles, and
Dallas. Every year, the center sponsors about a dozen trips to
the USSR, involving hundreds of people. The only function of
these trips is to get to know the Soviet people and to extend the
hand of friendship.

There are now 200-plus private-sector organizations in the
U.S. that are doing similar work. They are all learning that the
Soviet people are warm, generous, and deeply loyal, actually
more like Americans than other Europeans, with a great sense
of humor and a great love of family.

So what? We all know that the Soviets are human beings!
But do we? When we have contemplated the total destruction of
Soviet cities, did we really think of the children, the mothers,
the brothers? Helen Caldicott summed it up when she said,
"There are not American babies and Russian babies. There are
just *babies*."[4]

In summary, we find that the Soviets have abided by their
nuclear treaties, regardless of how belligerent they may have
been. They fear war, which explains their obsession with civil
defense. And they are fellow humans, whom we really have no
desire to destroy. Indeed, as John Donne might have said, to
destroy them would be to destroy ourselves.

Astronomers debate whether life exists elsewhere in the

universe. Some think it is very likely. Others feel that, if such life exists, it would have contacted us long ago; in the words of Enrico Fermi, "Where are they?" It is just possible that we are the *only* advanced life forms in our own galaxy of 100 billion stars, or possibly the only life in the entire universe of 100 billion galaxies. If so, we are indeed "God's children" and bear a tremendous moral imperative to survive. In a nuclear war we would be warring not just against the enemy, but against humankind, and against God.

The Psychology of War

The unleashed power of the atom has changed everything save our modes of thinking, and thus we drift toward unparalleled catastrophe.

ALBERT EINSTEIN, 1946[5]

There is a euphoric sense in America today that the Soviet military machine is no longer a threat. This is quite possibly true, but only time will tell. Neither we nor they have significantly reduced our nuclear arsenals. Why are we both hanging onto these weapons? Do we contemplate using them against each other? Do we contemplate using them against China? Surely, we don't need them to fight small countries.

Making war is pretty much a matter of mindset. Making peace is the same.

In World War II, we, the English, the French, and the Russians fought against tremendous initial odds to defeat, finally and conclusively, one of the most depraved governments of all time. In the Pacific, we almost singlehandedly repelled, and eventually conquered, another viciously ambitious government. We had a marvelous sense of accomplishment, of camaraderie with our allies, and of our moral superiority. We were, as we would say now, on a high. But in the process, *50 million people died*. We tend to forget that.

The next world war will kill *outright*, in the first few days, 140 million *Americans alone* (Figure 45). Most of the remaining Americans will not live for more than a year. Many will die in the first few weeks of exposure to radiation and cold. The rest will die of starvation and disease, their immune systems compromised and sanitation nonexistent.

Yet, we go on thinking about World War III as if it were just a bigger World War II. This is the kind of mentality that keeps concentrating on building ever bigger and better weapons, when we already have the ability to destroy any country with the nearly 200 missiles from a single nuclear submarine.

President Eisenhower, in his January 1961 farewell address to the American people, warned:

> In the councils of government, we must guard against the acquisition of unwarranted influence, whether sought or unsought, by the military-industrial complex. The potential for the disastrous rise of misplaced power exists and will persist.[6]

Make no mistake about it, one of the most powerful forces pushing the Star Wars concept is the military-industrial complex, which is not concerned that military expenditures do not strengthen our economy and actually weaken it. Then, what are they concerned with? The military is concerned with maintaining its own power. Industry is concerned with profits.

We must recognize that we have always overarmed ourselves with nuclear weapons. We have *always* had more strategic nuclear weapons than the USSR (Figure 46), in spite of all the talk about "windows of vulnerability" and the like. Furthermore, a 1986 report by the Pentagon stated that the United States leads the Soviet Union in virtually every basic technology that could affect military capabilities over the next 10 to 20 years.[7]

Can we stop this irrational plunge toward world destruction? Of course we can. We have already taken important steps

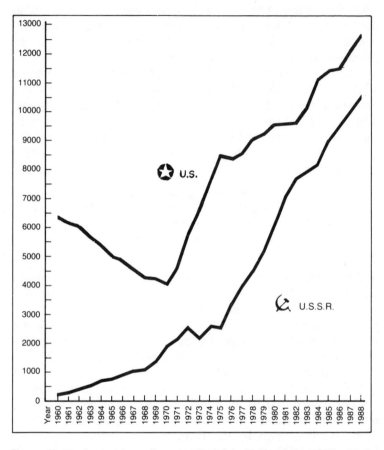

Figure 46. Total strategic nuclear weapons of the U.S. and the USSR over the period 1960–1988. (From *The Defense Monitor* **17**, No. 5, 1988. Courtesy of Center for Defense Information—see Bibliography.)

toward nuclear sanity. We did it most notably with the Limited Test Ban Treaty in 1963. We took another important step with the 1987 Intermediate-range Nuclear Forces (INF) Treaty, which removed destabilizing nuclear missiles from Europe. One of the easiest and most effective ways to reverse the nuclear arms race would be to ban all future testing of nuclear weapons—a comprehensive test ban. We need to press for this in the most serious possible way.

We have now integrated the nuclear weapon into the strategy of all of our armed forces. The system is becoming so complex—and Star Wars would make it more so—that it is becoming progressively more difficult to reverse, and the likelihood of accidental war is increasing. It is for reasons such as these that 6500 American physical scientists have refused to work on Star Wars projects. This is the same sector of the scientific community that eagerly worked on the Manhattan A-Bomb Project in World War II. Their opposition to this military initiative is unprecedented in the history of American science.

Scientists are not the only ones who are apprehensive. Thirty years ago, President Eisenhower said:

> I do not believe that I shall ever have to defend myself against the charge that I am indifferent to the fate of my countrymen. . . . When we get to the point, as we one day will, that both sides know that in any outbreak of general hostilities, regardless of the element of surprise, destruction will be both reciprocal and complete, possibly we will have sense enough to meet at the conference table with the understanding that the era of armaments has ended and the human race must conform its actions to this truth or die.[8]

There are many ways we can reduce the chance of nuclear war. But all of them require *a change of attitudes, a change of our perceptions of what war can accomplish*, and a recognition that we are dealing, not with arguments between superpowers, but with the destruction of the human race. We have to *stop* thinking the way we did during World War II.

We have to *start* thinking the way we shall have to think for the rest of human history. We have to grow up, put aside the war toys we have had so much fun with, and act like adults. We have to change our psychology, curb our fears, and open our minds and our hearts. If we yearn enough for peace, we shall find it.

Summary: No Nukes?

We have all seen, in our newspapers and on television, pictures of people carrying signs that say, "NO NUKES." What does that mean? Unfortunately, even the people who carry them appear not to know. Does it mean no saving of cancer patients? No cure for AIDS? No cheap, clean, nonpolluting power? Does it mean a unilateral destruction of all of our nuclear weapons?

Clearly, the sign is simplistic. It is carried by people who probably have not studied in any depth the problems of nuclear power and nuclear war. It is not easy to learn about these things, which is why I have written this book. I am not deriding those who carry such signs; they have a true and deep concern for the future of the human race. But in their ardor, they may unwittingly do more harm than good.

The Future of Nuclear Weapons

Is it realistic to press for the total elimination of nuclear weapons? Imagine such a future. Presumably, the outlawing of weapons would be policed by the International Atomic Energy Agency or the United Nations, which would prevent any large buildup of nuclear weaponry. One presumes that all nuclear testing would be prohibited, and that this prohibition would be monitored.

335

But we cannot erase the knowledge of how to make a nuclear bomb. It will be with us forever. If some nation acquired an insane leader, like Hitler, they might start making such weapons in the greatest secrecy, relying on their previous experience to be able to build bombs of some reliability without testing. They would have to make only a few weapons, perhaps a dozen. Theoretically, they could then hold at ransom any nation, or perhaps even the world.

Because of such scenarios, many thoughtful analysts conclude that it is unrealistic to attempt to ban nuclear weapons totally. However, it does seem reasonable, and probably essential, for us to reduce nuclear arsenals to a very small size. Suppose, for example, the U.S. and the USSR were permitted to have only 100 strategic nuclear warheads apiece, and that these would be restricted to submarines. Multiple-warhead (MIRVed) missiles would be forbidden, so, at about 20 missiles per sub, five nuclear submarines would be permitted. Only two major cities of each major nuclear power, presumably Washington and New York, and Moscow and Leningrad, would be allowed ABM defenses. France, Britain, and China would each be permitted only 40 warheads, on 2 submarines, and one ABM system. Nuclear weapons would be permitted to no other nation. The whole situation would, of course, by policed by an international agency, which would permit only rare, carefully observed tests. Such an arrangement would have the following benefits:

1. No nuclear nation would dare attack another, because of certain retaliation.
2. A counterforce attack would produce little radioactive fallout, as nuclear forces would be at sea.
3. An accidental release of a nuclear weapon would not damage the seat of government, because of its ABM defenses.
4. A non-nuclear nation could develop in secret no more than a few missiles, which the superpowers could oppose with a much larger and more sophisticated arsenal.

5. If an all-out exchange of weapons occurred, the damage would be low enough to preclude a nuclear winter.
6. The money saved by the superpowers would vastly aid their currently faltering economies.

I present this proposal merely as an example of what could be done. One would hope that something like this could provide a reasonable goal for arms-limitation negotiators, and it is essential in arms negotiation to have a feasible goal in mind. The day may come when we shall have a world government to which we can transfer all nuclear weapons. In the meantime, the size of present superpower nuclear arsenals is ridiculous, is horribly dangerous, and is destroying national economies. The ultimate absurdity is that no nation wants to use these arsenals because using them would be suicide.

"NO NUKES" *can apply largely* to the banning of nuclear weapons—but not completely.

The Future of Nuclear Power

Should a "NO NUKES" sign mean no nuclear power? It should be clear to those who have read this book that nuclear reactors can provide the world with clean and safe electric power.

However, people still question whether nuclear reactor technology is desirable. But just consider the major facts, all of which we have discussed:

1. The total number of people killed anywhere in the world in 35 years of nuclear power has been less than 50 (Chernobyl plus a few other small accidents with military and research reactors). The mining of coal in this century has killed over a million men in America alone.
2. There has been only *one* nuclear reactor disaster: Chernobyl.

3. Future cancer deaths from Chernobyl will be about $^1/_{3000}$ of those produced by smoking in the same population.
4. Nuclear waste can be safely disposed of.
5. Nuclear power will become economically competitive with oil-fueled power by 1995.
6. A normally operating nuclear power plant produces essentially no pollution, whereas power plants fired by coal and oil produce tremendous pollution.

Again, one is faced with the spectacle of people whose exaggerated fear of nuclear power blinds them to its true benefits. Nuclear power is dangerous. So is aviation. In fact, *flying is much more dangerous than nuclear power*. Hundreds of Americans die every year in aviation accidents. Yet we hear calls on all sides for a cessation of nuclear power because of one major accident. The closing of the Shoreham nuclear plant on Long Island is typical of the public's nuclear power hysteria.

A rational approach to nuclear power shows that all of the problems that this technology creates can not only be readily solved but have already been largely solved from a technical standpoint. Present reactors are very safe, with few exceptions, like the type used at Chernobyl. We know how to make inherently safe reactors that virtually cannot melt down, even if one tried to make them do so; the next generation of reactors will be of this type. We have good ideas about how to bury waste with complete safety, and another decade of the testing of proposed sites should perfect this technology.

The technical problems of nuclear power have been largely solved; the real block is political. One hears howls and screams when it is proposed to bury some nuclear waste deep in the ground, although the people howling and screaming will be exposed to far more radioactivity by the rocks directly underneath their houses. It is interesting that nuclear waste repositories are called "waste *dumps*" by the press. We do in fact dump chemical waste all over the place. But nuclear waste is never "dumped." Plans are to *encapsulate* and then *bury* nuclear waste *very deeply* in the ground. Such pejorative language is antagonistic to public understanding.

As for the economic problems, these are only temporary. In the very near future, nuclear power will once again be cost-competitive in the U.S. Even the low level of nuclear power that we have now decreases our dependence on foreign oil, a dependence that leads us to such extravagant responses as that evoked by the Iraqi takeover of Kuwait. Nuclear power also relieves environmental pollution. Many nations like France and Japan—both deeply committed to nuclear power—already recognize this.

"NO NUKES" *should not* apply to nuclear power.

The Future of Radioisotope Tools

I have discussed the use of radioisotopes in medicine and industry. Such use has marked a revolution of which the public is not generally aware.

Much of what we have learned in cell and molecular biology over the past 40 years has been due to the use of radioisotope tracer atoms and molecules. Thus, *much of our recent medical progress has resulted from the use of radioisotopes in research*. Our knowledge of cancer and of the diseases of the central nervous system, such as Alzheimer's disease and schizophrenia, has depended on this technology. The cure of AIDS requires extensive research using this technology. There is no way we can give up the use of radioisotopes in medical research.

Radioisotopes are also used in medical diagnosis and therapy. We *detect* some cancers with radioisotopes, and we *treat* many cancers with radiation from isotopes. Radioisotopes are also used widely in industrial and agricultural research and development. In short, the use of radioisotopes has permeated our modern society, and we can no more dispense with them than we can with computers.

There may be some concern about the disposal of radioactive wastes from these applications. However, except for some isotopes used in radiation therapy, the radioactivities of the isotopes used are almost always exceedingly low, and the half-lives are usually short. The tritium used in many biological experiments is at such a low level that it may be safely discarded down the sink. By the time it is diluted in the sewage system or in groundwater, its level of radioactivity is below that of naturally occurring isotopes. Any disposal problems related to the radioisotopes used in medicine and research are therefore minimal.

Anyone who says "NO NUKES" to the use of radioisotopes as tools in research and industry is saying no to cancer therapy and to AIDS research and to agricultural development.

"NO NUKES" *must not* apply to radioisotopes in medicine and industry.

Afterword: The Millennium

As you pass from sunlight into darkness and back again every hour and a half, you become startlingly aware how artificial are the thousands of boundaries we've created to separate and define. And for the first time in your life you feel in your gut the precious unity of the earth and all the living things it supports. The dissonance between this unity you see and the separateness of human groupings that you know exists is starkly apparent.

ASTRONAUT RUSTY SCHWEICKART
in Earth orbit; *Discover,* July 1987

We are approaching the year 2001, the beginning of the third millennium after Christ. It all began with a message of peace and brotherly love. By the end of the first millennium, the message had spread through the Western world, but the land was still full of misery and ignorance, pestilence and war. Then modern science began to function, slowly, a few seeds growing here and there in the fertile soil prepared 1500 years before by Aristotle.

Now, at the end of the second millennium, science and technology have come to full fruition, permeating all walks of life, curing misery, ignorance, and disease. But we have not cured war. Science and technology have not made war more likely, but they have certainly made it more destructive. However, the tools of technology can be turned to peaceful ends: the use of reconnaissance satellites actually decreases the chance of war by miscalculation. And the overwhelming presence of the nuclear bomb has frightened nations into drawing back several times from the precipice of war.

The curing of war cannot be accomplished by science and technology. It must be done by people: all of us. *Curing war will take all of the abilities that humans can muster.* It is quite evident that we cannot prevent war simply by eliminating nuclear weapons. Indeed, even to get those weapons down to low levels will require that we solve not just the problem of nuclear war, but the problem of *the concept of war*.

How do we do this? After all, Catholic fights Protestant in Ireland, and has done so, in one way or another, for centuries. Arab nation fights Arab nation. And it seems that almost everyone in Lebanon fights almost everyone else in Lebanon. One may wonder if there is any hope of peace.

There is hope. First of all, there is probably less war in the world today than at any time in the historical past. The United Nations, although a frequently ineffectual body, has nevertheless prevented many modern wars. Twenty-five wars are still going on in the world, but there were no new wars in the period 1980–1988.[1]

Remember, peace is a matter of mind-set. First of all, *we must believe that we can do it*. As President Eisenhower said:

> I like to believe that people in the long run are going to do more to promote peace than are governments. Indeed, I think that people want peace so much that one of these days governments had better get out of their way and let them have it.[2]

Is it impossible to imagine a world in which the Soviets and the Americans are bosom buddies? Is this a pipe dream? If you think so, just consider how most of us feel toward Japanese friends now, and how we felt about them right after Pearl Harbor. The same might be said about Americans and Iranians or Iraqis.

It seems inconceivable that California would ever contemplate going to war with New York. That it does not is rather remarkable: those two states are farther apart than London and Moscow. But peaceful attitudes can extend even farther: Does the United States ever contemplate war with Australia, halfway around the world? Clearly, distance alone does not prevent peaceful and friendly feelings. What brings these states and nations together in a peaceful and friendly rapport is primarily a common democratic conviction. It is less important that we share the same language, for to go to war with France would seem as strange as going to war with England. Many Americans are now visiting Russia, and they are experiencing some surprises. The Russian people are much like Americans: they share our pragmatism, our sense of humor. Why have we armed to the teeth against these nice people who are so much like us?

We may already be close to a *modus vivendi* with the Russians, a common ground of mutual respect and restraint. This could lead to extensive nuclear disarmament. But we have got to become convinced that we can have this peace. Such a conviction involves looking objectively at *our* behavior (such as knee-jerk responses to fictitious missile gaps) and looking objectively at *their* behavior (such as recognizing that they do keep treaties). In short, we must uncover the *myths* that separate us from the truth, and thus from peace in the world. We must learn that it is entirely possible for us to maintain a reasonable military stance and at the same time to provide for our poor and our disadvantaged, to sustain a sound economy, and to educate our children.

Second only to the problem of removing the scourge of war from humankind is the problem of controlling our destruction of the environment. Population control is, of course, a key element in the solution of both war and environmental destruction. But a big part of the latter problem is the burning of fossil fuels, which is altering our atmosphere in almost irreversible ways, through ozone destruction and the greenhouse effect. Gradual conversion to nuclear power, using the inherently safe reactors we now know how to make, would be a major rational step toward a cleaner environment. However, as in achieving nuclear disarmament, inertia and paranoia have prevented us from moving forward.

The release of the energy of the atomic nucleus will stand as one of the greatest achievements of humankind. We shall truly enter The Millennium if we use the energy of the nucleus to free the planet from pollution, and to free humankind from drudgery, rather than use it to destroy everything that civilization has constructed over the past millennia.

The nuclear lion remains fierce and can be easily provoked to destructiveness. But if we will only try, we can live easily with him and benefit from his awesome beauty and power.

⚛ Notes

Introduction

1. George F. Kennan, 1981, "On nuclear war." See *The Nuclear Delusion*, 1983. New York: Pantheon Books—Random House, pp. 194–195.
2. Carl Sagan, 1983, "To preserve a world graced by life," *Bulletin of the Atomic Scientists* **39** January, pp. 2–3.
3. In this book, simplifications are sometimes made that the purist may rightly question. They are meant to aid the lay reader. For example, categories of nuclear waste are defined in a quite different way than is current (see Chapter 9).

Chapter 2

1. Most textbooks state that organic molecules include all carbon compounds. But substances like carbon dioxide (CO_2), carbon disulfide (CS_2), and carbon tetrachloride (CCl_4) may be produced largely or solely by nonliving processes, whereas CH compounds are rarely so produced on Earth.
2. This statement is not exactly true. The enzymes work in concert only if they are in the proper environment, typically an existing biological cell.

Chapter 4

1. Air travel at 39,000 feet gives an enhanced cosmic-ray exposure of 0.0005 rem per hour (NCRP Report 93, September 1987—see Bibli-

ography). Thus, a round-trip New York-London flight exposes one to about an additional 0.005 rem of cosmic rays. For 100 such trips, this would be 0.5 rem per year.

2. "Health risks of radon and other internally deposited alpha-emitters," BEIR IV, 1988—see Bibliography; Marjorie Sun, 1988, "Radon's health risks," *Science* **239** (15 January), 250.

3. J. V. Neel, W. J. Schull, A. A. Awa, C. Satoh, H. Kato, M. Otake, and Y. Yoshimoto, 1990, "The children of parents exposed to atomic bombs: Estimates of the genetic doubling dose of radiation for humans," *American Journal of Human Genetics* **46** 1053–1072.

4. This is the *effective dose equivalent*, which is the biologically effective dose that is equivalent to the whole body being irradiated (see Appendix B). For example, an average chest X ray may involve a 0.015-rem dose to lung tissue, but this dose is equivalent in risk to only 0.006 rem of whole-body irradiation.

5. "Health effects of exposure to low levels of ionizing radiation," BEIR V, 1990—see Bibliography; Eliot Marshall, 1990, "Academy panel raises radiation risk estimate," *Science* **247** (5 January) 22–23.

6. NCRP Report 91, June 1987—see Bibliography. This limit is consistent with that of the International Commission on Radiological Protection (ICRP).

7. NCRP Report 93, September 1987—see Bibliography.

Chapter 5

1. Unfortunately, iodine-131 accumulates in the ovaries, but much less than in the thyroid. Although potentially hazardous for progeny, no leukemia or genetic abnormalities have been noted in the more than 30,000 patients treated so far—perhaps because of the high radiation resistance of those ovarian cells that produce eggs.

Chapter 6

1. The fission reaction with plutonium-239 is

$$^{239}_{94}\text{Pu} + ^{1}_{0}\text{n} \rightarrow {}^{240}_{94}\text{Pu} \rightarrow {}^{138}_{54}\text{Xe} + {}^{98}_{40}\text{Zr} + 4\,{}^{1}_{0}\text{n} + \text{gamma rays}$$

In practice, this reaction will release up to three neutrons, four being a theoretical upper limit.

2. The fusion reaction of deuterium with tritium is

$$^2_1H + {}^3_1H \rightarrow {}^4_2He + {}^1_0n + \text{gamma rays}$$

To avoid the problems of tritium's being a gas with a relatively short half-life (12 years), fusion bombs often use the solid compound *lithium deuteride* (LiD). Fission neutrons convert the lithium to helium plus tritium:

$$^6Li + {}^1n \rightarrow {}^4He + {}^3H$$

One then has both tritium and deuterium for the fusion process.

Chapter 7

1. Data on power-plant emissions are derived from Eisenbud, 1987; Lave and Freeburg, 1973; and UNSCEAR, 1988—see Bibliography. Also from J. P. McBride, R. E. Moore, J. P. Witherspoon and R. E. Blanco, 1978, "Radiological impact of airborne effluents of coal and nuclear plants," *Science* **202**, 1045–1050.
2. This is an average of the Nuclear Regulatory Commission regulations for nuclear power plant emissions of less than 0.005 rem per year from liquid discharges, as well as 0.010 rem per year of gamma radiation and 0.020 rem per year of beta radiation from gaseous emissions (Eisenbud, 1987, p. 213—see Bibliography).
3. Cohen, 1983, p. 186—see Bibliography. A more recent report states, "there has never been anyone injured radiologically or much less killed as a result of exposure to radiation or contamination arising as the result of an accident involving radioactive material during transportation" (Robert M. Jefferson, 1990, "Transporting radioactive materials and possible radiological consequences from accidents as might be seen by medical institutions," Chapter 3, in Fred A. Mettler, Jr., Charles A. Kelsey, and Robert C. Ricks, Eds., *Medical Management of Radiation Accidents*. Boca Raton, FL: CRC Press). See also Cohen, 1990—see Bibliography.
4. Cohen, 1990, pp. 240–246—see Bibliography.
5. ERDA Safeguards Program, Background Statement, 10 March 1975.

6. Eliot Marshall, 1986, "If terrorists go nuclear," *Science* **233** (11 July) 148–149.
7. See Note 1, Chapter 6.

Chapter 8

1. These and other reactor accidents are discussed by Cohen, 1990, Eisenbud, 1987, and Hall, 1984—see Bibliography.
2. Daniel F. Ford, 1982, *Three Mile Island: Thirty Minutes to Meltdown.* New York: Penguin Books. This book has an excellent summary and discussion of the events that occurred at Three Mile Island, but the conclusions are unduly pessimistic.
3. W. Booth, 1987, "Postmortem on Three Mile Island," *Science* **238**, 1342–1345.
4. Alvin M. Weinberg, 1989, "Science, government, and information: 1988 perspective," *Bulletin of the Medical Library Association* **77** (January), 1–7.
5. NCRP Report 93, September 1987—see Bibliography. The *maximum* individual dose outside the plant at Three Mile Island was less than 0.10 rem, equivalent to 4 months of background radiation.
6. Eisenbud, 1987, pp. 372–373—see Bibliography.
7. *PSR Newsletter*, 1983, **4** (4; Winter). Physicians for Social Responsibility, 1000 16th Street NW, Suite 810, Washington, DC 20036.
8. "Resurrecting a nuclear accident," 1983, *Nature* **302**, p. 207.
9. David Dickson, 1988, "Doctored report revives debate on 1957 mishap," *Science* **239**, 556–557.
10. An excellent review is the one by Mike Edwards, 1987, "Chernobyl—One year after," *National Geographic* **171** (May) 632–653. An authoritative summary appears in UNSCEAR, 1988—see Bibliography. See also J. Walsh, 1986 "Chernobyl: Errors and design flaws," *Science* **233**, 1029–1031; and J. F. Ahearne, 1987 "Nuclear power after Chernobyl," *Science* **236**, 673–679.
11. The reactions are complex. Graphite-moderated reactors get a small amount of their moderation from the water. Chernobyl is overmoderated, so loss of water, by conversion to steam, while decreasing moderation, can actually increase reactor power. Other factors also enter. See *Nuclear Safety: Comparison of DOE's Hanford N-Reactor*

with the Chernobyl Reactor, U.S. General Accounting Office, Briefing Report to Congressional Requesters, August 1986, p. 1—GAO/RCED86–213BR.

12. See C. Norman and D. Dickson, 1986, "The aftermath of Chernobyl," *Science* **233**, 1141–1143; also Eisenbud, 1987, and UNSCEAR, 1988—see Bibliography.

13. The Chernobyl power station is 4 miles east of the town of Pripyat (pop. 49,000) and 9 miles northwest of the town of Chernobyl (pop. 12,500).

14. See L. R. Anspaugh, R. J. Catlin, and M. Goldman, 1988, "The global impact of the Chernobyl reactor accident," *Science* **242** (16 December), 1513–1519.

15. W. Goodman, 1989, "Chernobyl revisited," *New York Times* (14 February).

16. For example, although heavy rains caused fallout in northwest England that was roughly comparable to that of the Windscale accident, levels elsewhere in Britain were very much lower (M. J. Clark and F. B. Smith 1988 *Nature* **322**, 245–249).

17. NCRP Report 93, September 1987—see Bibliography.

18. Peter Reizenstein, 1987, "Carcinogenicity of radiation doses caused by the Chernobyl fall-out in Sweden, and prevention of possible tumors," *Medical Oncology & Tumor Pharmacotherapy* **4**, 1–5.

19. Serge Schmemann, 1988, "Chernobyl and the Europeans: Radiation and doubts linger," *New York Times* (12 June).

20. Thyroid cancer and leukemia develop in 3 to 10 years. Many leukemias can now be cured, and thyroid cancer is highly curable with radioiodine.

21. Alvin M. Weinberg and Irving Spiewak, 1984, "Inherently safe reactors and a second nuclear era," *Science* **224**, 1398–1402.

Chapter 9

1. L. Roberts, 1982, "Ocean dumping of radioactive waste," *BioScience* **32**, 773–776; and Cottrell, 1981—see Bibliography.

2. Robert Leifer, Z. R. Juzdan, W. R. Kelly, J. D. Fassett and K. R. Eberhardt, 1987, "Detection of uranium from *Cosmos-1402* in the stratosphere," *Science* **238**, 512–514.

3. *FAS Public Interest Report* **41** (9; November 1988). See also **42** (9; November 1989). Federation of American Scientists, 307 Massachusetts Avenue NE, Washington, DC 20002.

4. An excellent discussion of the problems, both technical and political, of high-level nuclear waste disposal is presented by Carter, 1987—see Bibliography.

5. Bernard L. Cohen, 1977, "The disposal of radioactive wastes from fission reactors," *Scientific American* **236** (June), 21–31. See also Cohen, 1990—see Bibliography. Estimates of how much the radioactivity of waste fuel will drop in the first few years are difficult to make because the activity drops precipitously even in the first few days.

6. Cohen, 1983, p. 123; also Cohen, 1990—see Bibliography. See also Bernard L. Cohen, 1983, "Effects of recent neptunium studies on high-level waste hazard assessments," *Health Physics* **44**, 567–569.

7. ReVelle and ReVelle, 1984, p. 313—see Bibliography.

8. Colin Norman, 1982, "A long-term problem for the nuclear industry," *Science* **215**, 376–379.

9. James E. Campbell and Robert M. Cranwell, 1988, "Performance assessment of radioactive waste repositories," *Science* **239**, 1389–1392.

10. See Carter, 1987, and Milnes, 1985—see Bibliography.

11. Eliot Marshall, 1988, "Nevada wins the nuclear waste lottery," *Science* **239**, 15.

12. Carter, 1987, p.156—see Bibliography.

13. See, for example, S. Colley and J. Thomson, 1990, "Limited diffusion of U-series radionuclides at depth in deep-sea sediments," *Nature* **346** (19 July), 260–263.

Chapter 10

1. If you drove between two American cities during 1983–1985, you were 32 times more likely to be killed than if you made the same trip on a scheduled airliner. In addition, you were much more likely to have a nonlethal auto accident resulting in serious injuries. In 1986, there were *no deaths* on large scheduled airlines in the U.S. (*Accident Facts—1987* National Safety Council, Chicago).

2. Slovic, Fischhoff, and Lichtenstein, 1980—see Bibliography.

3. Cottrell, 1981—see Bibliography.
4. Carter, 1987—see Bibliography.
5. An official of Texas Utilities Electric, which built this plant, once debated with me that the plant was not upwind of Fort Worth, as the predominant surface winds are from the south; however, he failed to note the fact (of which all airplane pilots are aware) that winds shift to the west with altitude: at a few thousand feet, to which most plant emissions would rise, the predominant winds over Comanche Peak are from the southwest.
6. Matthew L. Wald, 1988, "At troubled nuclear plant, prepared for operation—or defeat," *New York Times* (1 May).
7. Mark Crawford, 1988, "Weapons legacy: A \$110-billion mess?" *Science* **241** (9 July), 155.
8. Klaus-Rudiger Trott and C. Streffer, 1990, "Occupational radiation carcinogenesis," Chapter 3, in E. Scherer, C. Streffer, and K-R. Trott (Eds.), *Radiation Exposure and Occupational Risks*. Berlin: Springer-Verlag.
9. Eliot Marshall, 1990, "Hanford releases released," *Science* **249** (3 August), 474.
10. R. Jeffrey Smith, 1982, "Atom bomb tests leave infamous legacy," *Science* **218**, 266–269; R. Jeffrey Smith, 1985, "NRC finds few risks for atomic vets," *Science* **228**, 1409.
11. Council on Scientific Affairs, American Medical Association, 1989, "Medical perspective on nuclear power," *Journal of the American Medical Association* **262**, 2724–2729.

Chapter 11

1. Frederick Seitz, 1990, "Must we have nuclear power?" *Reader's Digest* (August), 113–118.
2. David Sheridan, 1977, "A second coal age promises to slow our dependence on imported oil," *Smithsonian* **8** (August), 31–36.
3. Interview with Herschel H. Potter, former chief of coal mine health and safety of the Mine Safety and Health Administration, by New York Times correspondent Ben A. Franklin, 1988, *New York Times* (10 January).
4. Worldwatch Institute, *State of the World, 1987*. New York: W. W.

Norton, pp. 11, 12. In 1989, the U.S. Geological Survey stated that, at present rates of consumption, reserves and estimated undiscovered oil in the U.S. would last only 16 years. With 50% imports, it would last for 32 years: Richard A. Kerr, 1989, *Science* **245** (22 September), 1330–1331.

5. Alvin M. Weinberg, 1986, "Are breeder reactors still necessary?" *Science* **232**, 695–696.
6. Bernard L. Cohen, 1977, "The disposal of radioactive wastes from fission reactors," *Scientific American* **236** (June), 21–31.
7. "Nuclear electricity and energy independence," U.S. Council for Energy Awareness, 1776 I Street NW, Washington, DC 20006.
8. Allen A. Boraiko, 1985, "Storing up trouble. . . Hazardous waste," *National Geographic* (March), 318–351.
9. See Note 2, Chapter 6.
10. Robert Pool, 1989, "Cold fusion: End of Act I," *Science* **244**, 1039–1040.
11. J. Rafelski and S. E. Jones, 1987, "Cold nuclear fusion," *Scientific American* **257**, 84–89.
12. Kaku and Trainer, 1982, p.145—see Bibliography.
13. Philip H. Abelson, 1989, "Reliability of electric service," *Science* **245** (18 August), 689.
14. Lawrence Elliott, 1987, "Deadly winds: One year after Chernobyl," *Reader's Digest* (May), 129–133.
15. Richard Helms, 1990, interviewed in ". . . Resistance to blatant aggression, defense of allies and, above all, oil," *New York Times* (13 August).
16. Robert L. Hirsch, 1987, "Impending United States energy crisis," *Science* **235**, 1467–1473.

Chapter 12

1. U-238 fissions well with high-energy fast neutrons. However, the efficiency of the reaction is low compared to that achieved by U-235 with thermal (slow) neutrons, and it will not support a chain reaction; this is why U-235 is used in reactors and in fission bombs. U-238 can be used in the H-bomb blanket because a very high density of fast neutrons is available.

Chapter 13

1. Many of the subs in the fleets of both the U.S. and the USSR are older types, with less powerful weapons than the most modern subs (Table 5, footnote *b*). Therefore, the *equivalent* number of *modern* subs is less than the actual number of subs. By 1995, however, all the subs of both nations will be of the modern type.
2. Some analysts have suggested that such a response would be uncertain because communication with nuclear subs on patrol at great depths under the sea is slow and sometimes difficult. However, even though Washington might be knocked out quickly by an enemy nuclear sub attack, our military command centers far inland would have perhaps 20 minutes' warning before they were hit by ICBMs, during which time they could probably communicate with the subs; such command centers are presumably hardened against damage by electromagnetic pulse.
3. This section is based on material in Chivian, Chivian, Lifton, and Mack, 1982; and Peterson, 1983—see Bibliography.
4. Patricia J. Lindop and Joseph Rotblat, "The consequences of radioactive fallout," Chapter 18, in Chivian, Chivian, Lifton, and Mack, 1982—see Bibliography. A good discussion of fallout is presented by Charles S. Shapiro, T. F. Harvey, and K. R. Peterson, "Radioactive fallout," pp. 167–203, in Solomon and Marston, 1986—see Bibliography.
5. *PSR Reports* **6** (Summer 1985). Physicians for Social Responsibility, 1000 16th St. NW, Suite 810, Washington DC 20036.
6. Data from T. Ohkita, "Delayed medical effects at Hiroshima and Nagasaki," Chapter 7, in Chivian *et al.*, 1982, and updated by data from Shigematsu and Kagan, 1986—see Bibliography.
7. See, for example, pp. 224–225, in William Daugherty, Barbara Levi, and Frank von Hippel, "Casualties due to the blast, heat, and radioactive fallout from various hypothetical nuclear attacks on the United States," pp. 207–232, in Solomon and Marston, 1986—see Bibliography.

Chapter 14

1. Hiroo Kato, "Cancer mortality," in Shigematsu and Kagan, 1986—see Bibliography.

2. Pierce, 1989—see Bibliography.
3. Sohei Kondo, 1988, "Mutation and cancer in relation to the atomic-bomb radiation effects," *Japanese Journal of Cancer Research* **79**, 785–799.
4. BEIR V, 1990—see Bibliography.
5. J. V. Neel, W. J. Schull, A. A. Awa, C. Satoh, H. Kato, M. Otake, and Y. Yoshimoto, 1990, "The children of parents exposed to atomic bombs: Estimates of the genetic doubling dose of radiation for humans," *American Journal of Human Genetics* **46**, 1053–1072.
6. Sohei Kondo, 1988, "Altruistic cell suicide in relation to radiation hormesis," *International Journal of Radiation Biology* **53**, 95–102.
7. Information presented here is derived chiefly from Ehrlich, Sagan, Kennedy, and Roberts, 1984—see Bibliography, and from references cited therein.
8. Richard P. Turco, O. B. Toon, T. P. Ackerman, J. B. Pollack, and Carl Sagan, 1983, "Nuclear winter: Global consequences of multiple nuclear explosions," *Science* **222**, 1283–1292.
9. See pp. 128–130, in Mark A. Harwell and Christine C. Harwell, "Nuclear famine: The indirect effects of nuclear war," pp. 117–135, in Solomon and Marston, 1986—see Bibliography.
10. Around the year 1000, there was a warming trend throughout the Northern Hemisphere. The average temperature rose about 3° F, and the tree line moved 60 miles farther north! (Barry Lopez, 1986, *Arctic Dreams.* New York: Chas. Scribner's Sons, p. 184)
11. A 1985 report of a 2-year study by the Scientific Committee on Problems of the Environment (SCOPE) of the International Council of Scientific Unions, in which more than 300 scientists from 30 countries participated, estimated that only 1% of the world population could survive without organized agriculture. (Mark A. Harwell and Christine C. Harwell, "Nuclear famine: The indirect effects of nuclear war," pp. 117–135, in Solomon and Marston, 1986—see Bibliography.)
12. Alan Robock, 1989, "New models confirm nuclear winter," *Bulletin of the Atomic Scientists* **45** (September), 32–35.
13. Stephen H. Schneider, 1987, "Climate modeling," *Scientific American* **256** (May), 72–80.
14. Richard P. Turco, O. B. Toon, T. P. Ackerman, J. B. Pollack, and Carl Sagan, 1990, "Climate and smoke: An appraisal of nuclear winter," *Science* **247**, 166–176.

Chapter 15

1. Of the deep *body* waves, an earthquake produces mostly shear waves—waves that involve sliding one rock past another, thus cutting, or shearing, the rock layer. Explosions produce mostly compressional waves—waves that squeeze the rock or compress it. In addition, earthquakes produce a particular type of surface wave called a *Love wave*, which is not produced at all by explosions.
2. Richard A. Kerr, 1987, "U.S.-Soviet seismic monitoring advances," *Science* **235** (23 January), 434–435.
3. Scribner, Ralston, and Metz, 1985, p.1—see Bibliography.
4. R. Jeffrey Smith, 1983, "Scientists fault charges of Soviet cheating," *Science* **220** (13 May), 695–697.
5. R. Jeffrey Smith, 1987, "Alleged Soviet treaty violations rebutted," *The Washington Post* (13 February).
6. *The Defense Monitor* **16** (1; 1987). Center for Defense Information—see Bibliography.

Chapter 16

1. George W. Ball, 1985, "The war for Star Wars," *New York Review of Books* (11 April).
2. A favorable discussion is presented in J. E. Larson and W. C. Bodie, 1986, *The Intelligent Layperson's Guide to "Star Wars,"* National Strategy Information Center, 150 East 58 Street, New York, NY 10155. An opposing view is that of Tirman, 1986—see Bibliography.
3. Charles Mohr, 1986, "Delay in 'Star Wars' tests urged by 5 former U.S. policy makers," *New York Times* (17 August).
4. Colin Norman, 1988, "SDI deployment plan up in the air," *Science* **240** (17 June), 1608–1609.
5. There is no "top" to the atmosphere: it just gets thinner as you go up. At 15,000 feet, atmospheric pressure is $1/2$ that at the surface; at 30,000 feet, it is $1/4$, and so on. At 100,000 feet (20 miles), the air is too thin to support the flight of an airplane; one is above 99% of the atmosphere and can be regarded as being "above" it. At 200 miles, there is essentially no atmosphere; satellites at this altitude can, with no additional propulsion, simply coast in orbit around the Earth for many years.

6. AAAS annual meeting at Chicago, 16–20 February 1987, reported by Colin Norman, 1987, "The dark side of SDI," *Science* **235** (27 February), 962–963.

7. Mark Crawford, 1987, "SDI testing may ignite antisatellite race," *Science* **237** (31 July), 482.

8. Such an outcome is supported by a computer model created by Alvin Saperstein at Wayne State University, showing that when both countries introduced antimissile defense systems, the situation became chaotic and unstable, eventually leading to war (Robert Pool, 1989, "Everywhere you look, everything is chaotic," *Science* **245**, 7 July, 28).

9. See pp. 55–59 in Jonathan B. Tucker, "Scientists and Star Wars," Chapter 2, in Tirman, 1986, pp. 34–61—see Bibliography.

10. President's Commission on Strategic Forces (the Scowcroft Commission), final report from the Pentagon, 6 April 1983.

11. Documents are cited in P. B. Stares, 1985, *The Militarization of Space: U.S. Policy, 1945–84*, Cornell University Press, pp. 218–219.

12. William J. Broad, 1989, "Military to ready laser for testing as space weapon," *New York Times* (1 January).

13. John Tirman, "The politics of Star Wars," pp. 9, 10, 21, in Tirman, 1986—see Bibliography. Congress has consistently reduced administration funding requests for SDI. Expenditures for the first five years (1984–1988) were about $12 billion, instead of the $26 billion projected. In October 1990, Congress approved only $2.9 billion of the President's request of $4.7 billion for SDI.

Chapter 17

1. McGeorge Bundy, 1990, "From cold war toward trusting peace," *Foreign Affairs* **69**, 197–212.

2. Scribner, Ralston, and Metz, 1985, p. 10—see Bibliography.

3. Lynn R. Sykes and Jack F. Evernden, 1982, "The verification of a comprehensive nuclear test ban," *Scientific American* **247** (October), 47–55.

4. Paul Lewis, 1989, "Nonaligned nations seek total nuclear test ban," *New York Times* (15 November).

5. "The Freeze," Fact Sheet 2, 1983, Nuclear Weapons Freeze Campaign, now Sane/Freeze, Campaign for Global Security, 711 G Street SE, Washington, DC 20003.
6. Weinberg and Barkenbus, 1988—see Bibliography.

Chapter 18

1. M. Anderson, J. Brugmann, and G. Erickcek, 1983, *Destructive Investment, Nuclear Weapons and Economic Decline*, p. 1, and 1982, *The Price of the Pentagon*, p. 2. Lansing, MI: Employment Research Associates.
2. R. DeGrasse, 1983, *Military Expansion, Economic Decline*, Chapter 1. New York: Council on Economic Priorities.
3. Michael Dee Oden, 1988, "Military spending erodes real national security," *Bulletin of the Atomic Scientists* **44** (June), 36–42.
4. Dumas, 1986, p. 211—see Bibliography.
5. Colin Norman, 1990, "Defense research after the cold war," *Science* **247** (19 January), 272–273.
6. *WAND Bulletin* **9** (1; Winter 1990). Women's Action for Nuclear Disarmament, 691 Massachusetts Avenue, Arlington, MA 02174.
7. T. B. Edsall, 1988, "The return of inequality," *The Atlantic* (June).
8. Barbara J. Culliton, 1988, "It matters how you slice the pie," *Science* **240** (1 April), 19.
9. Paul Kennedy, 1987, *The Rise and Fall of the Great Powers*. New York: Random House.
10. *The Defense Monitor* **16** (3; 1987). Center for Defense Information—see Bibliography.
11. Mark Crawford, 1987, "Soviets interested in study on economic conversion," *Science* **235**, 1133.

Chapter 19

1. Richard Wakeford, 1988, "Incidence of leukemia," *Nature* **331**, 296 (correspondence).

Chapter 20

1. Helen Caldicott, 1986, *Missile Envy*. New York: Bantam Books.
2. I. Shigematsu and S. Akiba, "Sampling of atomic bomb survivors and method of cancer detection in Hiroshima and Nagasaki," in Shigematsu and Kagan, 1986, pp. 1–8—see Bibliography.
3. 3268 Sacramento Street, San Francisco CA 94115.
4. Helen Caldicott, in the videotape *The Last Epidemic*, produced and distributed by Physicians for Social Responsibility, 1000 16th Street NW, Suite 810, Washington DC 20036.
5. *Nuclear War Quotations*, p. 7. Center for Defense Information—see Bibliography.
6. Dwight D. Eisenhower, 1961, "Farewell radio and television address to the American people" (17 January). Item 421, *Public Papers of the Presidents*, pp. 1035–1040. Dwight D. Eisenhower Library, Abilene, KS 67410.
7. R. Jeffrey Smith, 1986, "U.S. tops Soviets in key weapons technology," *Science* **231**, 1063–1064.
8. Dwight D. Eisenhower, in a letter of 4 April 1956 to Richard L. Simon, president of Simon & Schuster, publishers. Reported by David S. Broder, 1983, ". . . We are rapidly getting to the point that no war can be won," *Washington Post* (7 September).

Afterword

1. M. J. Berlin, 1987, "P.S. Peace may be breaking out," *The Interdependent* **13** (5; October-November). New York: The United Nations Association.
2. Dwight D. Eisenhower, television talk with British Prime Minister Harold Macmillan, 6 September 1959. *Nuclear War Quotations*, p. 87. Center for Defense Information—see Bibliography.

Appendixes

A Bottom Part of the Periodic Table

This chart shows the elements in the periodic table from cesium to californium. There is a gap between lanthanum and hafnium (atomic numbers 58 to 71), represented by a group of "rare-earth" elements, which are not shown, that "loop out" from the periodic table. The elements beyond actinium line up with those rare earths and thus also form a "loop-out." Neptunium to californium are transuranic elements.

Mass	Symbol	Atomic No.	Name
133	Cs	55	Cesium
137	Ba	56	Barium
139	La	57	Lanthanum
178	Hf	72	Hafnium
181	Ta	73	Tantalum
184	W	74	Tungsten
186	Re	75	Rhenium
190	Os	76	Osmium
192	Ir	77	Iridium
195	Pt	78	Platinum
197	Au	79	Gold
201	Hg	80	Mercury
204	Tl	81	Thallium
207	Pb	82	Lead
209	Bi	83	Bismuth
210	Po	84	Polonium
210	At	85	Astatine
222	Rn	86	Radon
223	Fr	87	Francium
226	Ra	88	Radium
227	Ac	89	Actinium
232	Th	90	Thorium
231	Pa	91	Protactinium
238	U	92	Uranium
237	Np	93	Neptunium
242	Pu	94	Plutonium
243	Am	95	Americium
247	Cm	96	Curium
247	Bk	97	Berkelium
251	Cf	98	Californium

B Dosimetry

Absorbed Dose

The common unit of *absorbed dose* is the **rad,**[a] which represents the absorption of 1/100 of a joule of energy by 1 kilogram of the absorbing material. Because raising the temperature of 1 kilogram of water by 1 degree Celsius requires 4200 joules of energy, it can be seen that 1 rad ($^1/_{100}$ joule per kilogram) represents a tiny amount of energy.

Dose Equivalent (Biological Effectiveness of a Dose)

The common unit that measures the *biological effectiveness* of radiation *for the tissue irradiated* is the **rem,**[a] which is the amount of radiation that produces the same biological effect as is produced by the absorption of one rad of X-ray or gamma-ray energy. Thus,

$$\textbf{1 rad} = \textbf{1 rem} \quad (\text{X or gamma rays})$$

---

[a]The units *rad* and *rem* have been replaced by new international units:

1 gray (Gy) = 100 rads; 1 sievert (Sv) = 100 rems

I do not use these units in this book, because they are rather more difficult to use and because most of the literature is still in the older units.

Particle radiation generally produces a greater biological effect with the same energy. Consequently, rems usually do not equal rads for these particles. *Roughly speaking,*

$$\textbf{1 rad} \cong \textbf{1 rem} \quad \text{(beta rays, electrons)}$$
$$\textbf{1 rad} \cong \textbf{10 rems}^b \quad \text{(neutrons, protons, alpha particles)}$$

where $\cong$ means "is roughly equal to."

The whole-body lethal dose for a human is about 450 rems. The background dose received by the average American is about $3/10$ of a rem per year, or 21 rems in a 70-year lifetime.

Effective Dose Equivalent

The *effective* dose equivalent measures the *biological effectiveness* of a dose that is *equivalent to the whole body being irradiated*. It is the "dose equivalent" (in rems) multiplied by a weighting factor. It provides a measure of radiation-induced risk (somatic and genetic) to the individual, even though the body is *not* uniformly irradiated, and it makes it possible to compare, for example, the risk due to a chest X ray (which involves only part of the body) with the risk due to cosmic radiation (which involves the whole body).

Rate of Decay of a Radioisotope (Activity)

How radioactive a substance is can be measured in terms of the *rate of decay* of its *radioactive* atoms (many of the atoms may not be radioactive). The unit is called a **curie**[c] (Ci):

---

[b]This number depends on the dose, as the biological effect of these particles is roughly linear with dose, whereas the biological effect of X and gamma rays increases with dose (a sublinear effect). Thus, for fast neutrons, 1 rad $\cong$ 15 rems in the dose range 30–100 rads, whereas 1 rad $\cong$ 7 rems in the range 100–250 rads.

[c]The new international unit is the becquerel (Bq), which is one decay per second. Thus, 1 Ci = 37,000,000,000 Bq.

1 Ci = 37,000,000,000 decays per second

This curious number arose from the fact that *1 gram of pure radium-226 has an activity of 1 curie*. Thus, 37 billion of the radium atoms will decay every second. This number is a tiny fraction of the total number of atoms in a gram of radium, which is 3 *billion trillion*, or 3,000,000,000,000,000,000,000,000 atoms.

The **half-life** of a radioisotope is the time required for one-half of its atoms to decay. The half-life of radium-226 is 1620 years. In this time, 1.5 billion trillion of the atoms in 1 gram of pure radium will decay.[d]

[d]1620 years = 1620 × 365 × 24 = 14.2 million hours = 14.2 × 60 × 60 = 51 billion seconds. Therefore, 1 gram of radium will show 1.5 billion trillion decays in 51 billion seconds = *30 billion decays per second*. This calculation illustrates how it is that even those isotopes with very long half-lives, such as uranium-238 (5 billion years), will still show easily measurable radioactivity.

C   Important Radioisotopes Produced by Nuclear Reactors

Element	Isotope	Half-life	Biological hazard		
			Ingestion	Inhalation	Target
Neutron bombardment products[a]					
Tritium (g)[b]	^{3}H	12 years	x	x	All tissues
Carbon	^{14}C	5600 years	x	x	All tissues
Argon (g)	^{41}Ar	2 hours		x	Lungs
Manganese	^{54}Mn	1 year	x		All tissues
Iron	^{55}Fe	3 years	x		Hemoglobin
Cobalt	^{60}Co	5 years	x		GI tract
Niobium	^{94}Nb	20,000 years	x		Bone
Fission products					
Krypton (g)	^{85}Kr	11 years		x	Lungs
	^{88}Kr	3 hours			
Strontium	^{90}Sr	30 years	x	x	Bone (mimics Ca)
Iodine	^{129}I	17,000,000 years	x		Thyroid
	^{131}I	8 days		x	
Xenon (g)	^{133}Xe	5 days		x	Lungs
Cesium	^{134}Cs	2 years		x	All tissues
	^{137}Cs	30 years			(mimics K)

(continued)

Element	Isotope	Half-life	Biological hazard		
			Ingestion	Inhalation	Target
Uranics^c					
Lead	^{210}Pb	22 years	x		Liver
Radon	^{222}Rn	4 days		x	Lungs
Thorium	^{229}Th	7000 years	x		Bone
Transuranics^d					
Neptunium	^{237}Np	2,000,000 years	x		⎫ All tissues,
Plutonium	^{238}Pu	90 years		x	esp. liver
	^{239}Pu	24,000 years	x	x	Later, bone
	^{240}Pu	7000 years	x	x	(decay to
Americium	^{241}Am	500 years	x	x	Ra—mim-
	^{243}Am	8000 years	x	x	⎭ ics Ca)

[a] Produced by bombardment by neutrons released during fission. Tritium is produced in both air and water, argon in air. Manganese, iron, cobalt, and niobium radioisotopes are produced in the steel structures of the reactor core. The neutron bombardment products and the fission products are primarily beta-particle emitters, although Mn-54, Co-60, Nb-94, I-131, and both cesium isotopes also emit powerful gamma rays.

[b] All products are solids, except where noted as gases (g). Solids may become airborne as tiny particles or dissolved in water droplets.

[c] The uranium fuel is not listed. The isotopes shown, decay products of uranium and plutonium, emerge very late during storage of nuclear waste. The decay of plutonium-241 gives rise to important transuranics and uranics:

$$^{241}\text{Pu} \longrightarrow {}^{241}\text{Am} \longrightarrow {}^{237}\text{Np} \longrightarrow {}^{233}\text{U} \longrightarrow {}^{229}\text{Th}$$
h.l. (yrs)   13   500   2,000,000   16,000

[d] These are alpha-particle emitters, generally emitting only weak gamma rays.

Sources: Bernard L. Cohen, 1982, "Effects of ICRP Publication 30 and the 1980 BEIR report on hazard assessments of high-level waste," *Health Physics* **42,** 133–143; Ronnie D. Lipschutz, 1980, *Radioactive Waste: Politics, Technology, and Risk.* Cambridge, MA: Ballinger; Milnes, 1985, and UNSCEAR, 1977—see Bibliography.

Bibliography

This is a limited bibliography, mostly of books that summarize technical subjects in a way that can be readily understood by the layperson. The asterisk (*) at the left margin indicates books that are particularly recommended.

Among many groups that specialize in information about nuclear matters, two of the most prominent and reliable are the following, whose data are often cited in this book:

Union of Concerned Scientists
26 Church Street
Cambridge, MA 02238   (617-547-5552)

A nonprofit organization of scientists and other citizens concerned about the impact of advanced technology on society, including nuclear arms limitation, energy policy alternatives, and nuclear power safety. Organized by scientists at MIT, its present Chairman of the Board is Henry W. Kendall, a professor of physics at MIT. Quarterly publication: *Nucleus*.

Center for Defense Information
1500 Massachusetts Avenue NW
Washington, DC 20005   (202-862-0700)

A nonprofit, nonpartisan research organization, founded in 1972 and directed by retired officers of the U.S. military. Provides accurate information on, and appraisals of, the military weapons and arsenals of the world powers. Founder and director is Rear Admiral Gene R. LaRocque, U.S. Navy (Ret.). Monthly publication: *The Defense Monitor*.

BEIR IV, 1988, *Health Risks of Radon and Other Internally Deposited Alpha-Emitters*. Committee on the Biological Effects of Ionizing Radiations, National Research Council. Washington, DC: National Academy Press.

BEIR V, 1990, *Health Effects of Exposure to Low Levels of Ionizing Radiation*. Committee on the Biological Effects of Ionizing Radiations, National Research Council. Washington, DC: National Academy Press.

Bundy, McGeorge, 1990, "From Cold War toward trusting peace." *Foreign Affairs* **69**, 197–212.

*Carter, L. J., 1987, *Nuclear Imperatives and Public Trust: Dealing with Radioactive Waste*. Resources for the Future, Washington, DC. Baltimore, MD: Johns Hopkins University Press. $28.

*Chivian, E., S. Chivian, R. J. Lifton, and J. E. Mack, 1982, *Last Aid: The Medical Dimensions of Nuclear War*. San Francisco: W.H. Freeman. $10. paper.

Cohen, B. L., 1983, *Before It's Too Late: A Scientist's Case FOR Nuclear Energy*. New York: Plenum Press. $17.

*Cohen, B. L., 1990, *The Nuclear Energy Option: An Alternative for the 90s*. New York: Plenum Press. $25.

*Cottrell, A., 1981, *How Safe Is Nuclear Energy?* London: Heinemann. $6. paper.

Deutsch, R. W., 1987, *Nuclear Power: A Rational Approach* (4th ed.). Columbia, MD: GP Courseware. $5. paper.

Duderstadt, J. J., 1979, *Nuclear Power*. New York: Marcel Dekker.

*Dumas, L. J., 1986, *The Overburdened Economy*. Berkeley, CA: University of California Press. $19.

Edwards, M., 1987, "Chernobyl—one year after," *National Geographic* **171** (May), 632–653.

*Ehrlich, P. R., C. Sagan, D. Kennedy, and W. O. Roberts, 1984, *The Cold and the Dark: The World After Nuclear War*. New York: W.W. Norton. $14.

*Eisenbud, M., 1987, *Environmental Radioactivity* (3rd ed.). Orlando, FL: Academic Press. $57.

Gorbachev, M., 1987, *Perestroika*. New York: Harper & Row. $20.

Haley, P. E., and J. Merritt, 1988, *Nuclear Strategy, Arms Control, and the Future*. Boulder, CO: Westview Press. $17. paper.

*Hall, E. J., 1984, *Radiation and Life* (2nd ed.). Elmsford, NY: Pergamon Press. $20.

Kaku, M., and J. Trainer, 1982, *Nuclear Power: Both Sides*. New York: W.W. Norton. $7. paper.

*Kennan, G. F., 1983, *The Nuclear Delusion—Soviet–American Relations in the Atomic Age*. New York: Pantheon. $6. paper.

Lapp, R. E., and H. L. Andrews, 1963, *Nuclear Radiation Physics* (3rd ed.). Englewood Cliffs, NJ: Prentice-Hall.

Lave, L. B., and L. C. Freeburg, 1973, "Health effects of electricity generation from coal, oil, and nuclear fuel," *Nuclear Safety* **14**, 409–428.

Lester, R. K., 1986, "Rethinking nuclear power," *Scientific American* **254** (March), 31–39.

Lipschutz, R. D., 1980, *Radioactive Waste: Politics, Technology, and Risk*. Cambridge, MA: Ballinger (Harper & Row).

*Milnes, A. G., 1985, *Geology and Radwaste*. New York: Academic Press. $40. paper.

Mould, R. F., 1988, *Chernobyl: The Real Story*. Elmsford, NY: Pergamon Press. $18. paper.

Murray, R. L., 1988, *Nuclear Energy*. Elmsford, NY: Pergamon Press. $25. paper.

NCRP Report #91, 1987 (June), *Recommendations on Limits for Exposure to Ionizing Radiation*. National Council on Radiation Protection and Measurements, 7910 Woodmont Avenue, Bethesda, MD 20814.

NCRP Report #93, 1987 (September), *Ionizing Radiation Exposure of the Population of the United States*. National Council on Radiation Protection and Measurements, 7910 Woodmont Avenue, Bethesda, MD 20814.

*Peterson, J. (Ed.), 1983, *The Aftermath: The Human and Ecological Consequences of Nuclear War*. Royal Swedish Academy of Sciences. New York: Pantheon. $7. paper.

Pierce, D. A., 1989, "An overview of the cancer mortality data on the atomic bomb survivors," RERF CR 1–89. Hiroshima, Japan: Radiation Effects Research Foundation.

Ramsay, W., 1979, *Unpaid Costs of Electrical Energy*. Resources for the Future, Washington, DC. Baltimore, MD: The Johns Hopkins University Press. $10. paper.

ReVelle, P., and C. ReVelle, 1984, *The Environment: Issues and Choices for Society* (2nd ed.). Boston: Willard Grant Press. $28.

*Roller, A., 1974, *Discovering the Basis of Life*. New York: McGraw-Hill. $22. paper.

*Schell, J., 1982, *The Fate of the Earth*. New York: Avon Books. $3. paper.

*Scribner, R. A., T. J. Ralston, and W. D. Metz, 1985, *The Verification Challenge*. American Association for the Advancement of Science Project. Boston: Birkhauser. $19. paper.

Shigematsu, I., and A. Kagan, 1986, *Cancer in Atomic Bomb Survivors*. Tokyo: Japan Scientific Societies Press; and New York: Plenum Press. $55.

Shimizu, Y., H. Kato, and W. J. Schull, 1988, "Life Span Study Report 11. Part 2. Cancer mortality in the years 1950-85 based on the recently revised doses (DS86)," RERF TR 5–88. Hiroshima, Japan: Radiation Effects Research Foundation.

Slovic, P., B. Fischhoff, and S. Lichtenstein, 1980, "Facts and fears: Understanding perceived risk," in *Societal Risk Assessment* (R.C. Schwing and W. A. Albers, Jr., Eds.). New York: Plenum Press, pp. 181–216.

Solomon, F. and R. Q. Marston (Eds.), 1986, *The Medical Implications of Nuclear War*. Washington, DC: National Academy Press. $34. paper.

Sweet, W., 1984, *The Nuclear Age: Power, Proliferation and the Arms Race*. Washington, DC: Congressional Quarterly. $13. paper.

Talbott, S., 1985, *Deadly Gambits*. New York: Vintage Books/Random House. $8. paper.

*Tirman, J. (Ed.), 1986, *Empty Promise: The Growing Case Against Star Wars*. Union of Concerned Scientists. Boston: Beacon Press. $8. paper.

UNSCEAR, 1988, *Sources, Effects and Risks of Ionizing Radiation*. United Nations Scientific Committee on the Effects of Atomic Radiation, 1988 Report to the General Assembly, with annexes. New York: United Nations. $90.

Upton, A. C., 1982, "The biological effects of low-level ionizing radiation," *Scientific American* **246**, 41–49.

*Wagner, H. N., Jr., and L. E. Ketchum, 1989, *Living with Radiation: The Risk, the Promise*. Baltimore, MD: Johns Hopkins University Press. $18.

Weinberg, A. M., and J. N. Barkenbus (Eds.), 1988, *Strategic Defenses and Arms Control*. New York: Paragon House. $25.

Zeckhauser, R. J., and W. K. Viscusi, 1990, "Risk within reason," *Science* **248**, 559–564.

⚛ Glossary

All items in this glossary are shown in **boldface** the first time they appear in the text. Boldface items within a definition occur elsewhere in the glossary.

activity   The degree of **radioactivity** of a substance.

AEC   Atomic Energy Commission. Agency in charge of all national atomic (nuclear) energy matters, both civilian and military, from 1946 to 1975. Superseded by **DOE** and **NRC**.

alpha particle (ray)   A subatomic particle, identical to the helium nucleus, consisting of two **protons** and two **neutrons**, and thus bearing a double positive electric charge. One of the radiations emitted by a **radioisotope**. A poor penetrator of matter, being stopped by a piece of cardboard.

artificial radioisotope   A **radioisotope** produced in the laboratory, as by bombardment with **subatomic particles**.

atom   Elementary unit of matter. Also called an **element**. There are 92 different natural atoms in the universe. All material things are made of atoms, including **molecules**. Atoms have a tiny **nucleus**, consisting of **protons** and **neutrons**, around which **electrons** travel in orbits, much as planets around the Sun. Examples: hydrogen, iron, oxygen.

atomic bomb (A-bomb)   A **fission bomb**.

atomic number   The number of **protons** in the nucleus of an atom, varying from 1 in hydrogen to 92 in uranium. This number determines the chemical properties of the atom.

atomic weight   The sum of the numbers of **protons** and **neutrons** in the nucleus of an atom, varying from 1 in hydrogen to 238 in common uranium (uranium-238). The atomic weight characterizes the weight of the atom, as the electrons are essentially weightless.

ballistic missile A rocket-fired missile that shuts off its rocket early and then coasts throughout the rest of its trajectory. Called *ballistic* because its flight is similar to that of a shell fired from a cannon.

beta particle (ray) A subatomic particle, identical to the electron, emitted from the nucleus of a **radioisotope**. A poor penetrator of matter, being stopped by a pane of glass.

blanket In a **fusion** (hydrogen) **bomb**, the outer layer of uranium-238 that fissions from the fast neutrons produced in the fusion reaction.

blast wave A spherical wave, traveling at the speed of sound, that spreads out from a nuclear explosion. It behaves like an extremely strong and sudden wind; it can shatter buildings and kill people.

boiling-water reactor (BWR) A water-cooled **nuclear reactor** in which the water is permitted to boil as it passes through the reactor core. The steam produced is used directly to turn a turbine, which is connected to an electric generator.

boost phase The period in the trajectory of a **ballistic missile** during which the main launch rockets are firing. For an **ICBM**, about 3 to 5 minutes.

breeder reactor A **nuclear reactor** that produces more fissionable fuel than it burns. Typically, such a reactor burns plutonium-239 and uses the neutrons produced to convert uranium-238 into more plutonium-239.

cell The unit of a living organism. All the substance of our bodies consists of cells or the products of cells. Living cells have a **nucleus** surrounded by cytoplasm.

chaff Metal wires or strips, such as aluminum foil, deployed by aircraft or ballistic missiles to confuse enemy radar.

chain reaction A reaction that grows geometrically, resulting in explosion. Ordinary (TNT) bombs undergo *chemical* chain reactions, but the term is especially applied to the *nuclear* chain reaction.

chemical bond The force that holds atoms tightly together, to form a **molecule**.

chromosome A long, thin structure in the **nucleus** of a living cell that is easily stained for microscopic examination (whence its name, from the Greek, meaning "colored body"). It contains **DNA**, which is the genetic material of the cell. Human cells have 23 pairs of chromosomes.

cladding The hollow **fuel rods** that surround the pellets of uranium fuel used in a **nuclear reactor**. Usually made of zirconium alloy or stainless steel.

comprehensive test ban A proposed treaty that would ban all tests of nuclear weapons. A comprehensive test ban is not as sweeping as a **nuclear freeze**.

confinement Long-term safe storage of **radwaste** in a manner that permits retrieval.

control rod Rod made of a metal that absorbs neutrons very well, that can be raised and lowered within a nuclear **reactor core** to control the reaction. Made of the elements boron, cadmium, or hafnium.

conversion reaction A reaction in which one nuclear fuel is converted into another by the capture of neutrons. The term is applied particularly to the conversion of uranium-238 to plutonium-239, which occurs in all uranium reactors, but is especially utilized in a **breeder reactor**.

counterforce Relating to weapons or strategy aimed at destroying the offensive nuclear capability of the enemy, represented by missile silos, air bases, submarines, and command centers.

countervalue Relating to weapons or strategy aimed at destroying enemy cities and population.

critical condition The state of the core of a nuclear reactor when it is maintaining a controlled chain reaction.

critical mass The minimum mass of fissionable material required to sustain a chain reaction. Actually a slight misnomer, as the *shape*, as well as the total mass, is involved.

cruise missile A "flying bomb," like the V-1 weapons of World War II. A missile with a jet engine that flies at high, but subsonic, speed and very low altitude. Generally capable of following a map electronically and changing course during flight, so as to actively "seek out" its target.

curie The unit that measures the rate of decay (**activity**) of a **radioisotope**. Roughly, the radioactivity of 1 gram of pure radium (see Appendix B).

daughter isotope The **isotope** resulting from radioactive decay. Thus, uranium-238 decays through **alpha-particle** emission to the daughter isotope thorium-234. The daughter isotope will always be a chemically different atom (having a different **atomic number**) from the parent isotope.

decay product   Daughter isotope.

decoy warheads   Fake warheads, of light weight, that may be discharged by an **ICBM** in midcourse to confuse enemy defenses. Unlike the real warhead, they burn up on reentry into the atmosphere.

deuterium   A heavy **isotope** of hydrogen, with one proton and one neutron in the nucleus. One of every 5000 atoms of hydrogen in natural waters is deuterium.

directed-energy weapon   An antiballistic missile, planned for a **Star Wars** defense, that utilizes a powerful beam of light from a laser, or a beam of atoms or **subatomic particles** from an electric gun.

DNA   *D*eoxyribo*n*ucleic *a*cid. The genetic material of all living organisms, residing in the **chromosomes** of the cell. It is a long, threadlike molecule, normally occurring as a double helix of two intertwined DNA strands, each a long chain of nucleotides. Each nucleotide contains a base, either adenine (*A*), guanine (*G*), thymine (*T*), or cytosine (*C*). The sequence of these bases provides a genetic message that determines the amino-acid sequence of **proteins**, which in turn provide the essential physical structures and chemical functions of the cell.

DOE   U.S. Department of Energy, established in 1977. Performs many of the functions of the former Atomic Energy Commission (**AEC**).

dose   Usually, the amount of radiation absorbed by an object. Usually measured in *rads* or **rems** (see Appendix B).

electromagnetic pulse   A radio wave of extremely high intensity (high voltage gradient), which can destroy unshielded electrical and electronic equipment. Produced by the **gamma rays** from a nuclear explosion in space, which strike the atmosphere and ionize its atoms.

electromagnetic radiation   Radiation consisting of units called **photons**, which have associated vibrating electric and magnetic fields, have no **mass**, and travel at the speed of light. Examples are (in decreasing order of energy) **gamma rays, X rays**, visible light, microwaves, and radio waves. See Figure 5.

electromagnetic rail gun   An antiballistic-missile weapon, planned for a **Star Wars** defense, that accelerates a very small metal projectile to extremely high velocities by passing it down a tube through strong electric fields.

electron A subatomic particle with a unit negative electric charge that orbits around the positively charged nucleus of the atom. In an electrically neutral (normal) atom, the number of (negative) electrons equals the number of (positive) protons.

element A substance made of a single kind of **atom**, such as iron or iodine. Characterized by its **atomic number**.

energy The ability to do work.

enzyme A globular **protein** that acts as a catalyst of chemical reactions in living systems. Almost all chemical reactions in living cells are carried out by enzymes; without the enzymes, these reactions could occur only at elevated temperatures inimical to the living state. The enzyme *pepsin* in the stomach digests protein.

EPA Environmental Protection Agency, founded in 1970. Federal agency in charge of monitoring all aspects of environmental pollution and of enforcing laws to clean up the environment.

fallout The radioactive dust and debris that fall to the ground after a nuclear explosion.

firestorm A devastating fire produced by powerful convection winds that are induced by the extreme heat of a nuclear explosion or of a large number of conventional incendiary bombs. The firestorm stokes existing fires, leading to total conflagration.

first strike The strategy of striking first with nuclear weapons at an enemy suspected to be about to begin a war.

fission bomb (atomic bomb) A bomb whose explosion results from nuclear fission of uranium-235 or plutonium-239.

fission product A daughter atom from **nuclear fission**. Uranium-235, for example, may split to produce the fission products xenon and strontium.

fossil fuel Fuel produced by the slow underground compression and chemical change of plants that existed over 100 million years ago. Includes coal, oil, and natural gas.

fuel rod A tube that contains the fissionable fuel pellets in a nuclear **reactor core**. See **cladding**.

fusion bomb (hydrogen bomb) A bomb whose explosion results largely from nuclear fusion of hydrogen **isotopes**. Uses a **fission bomb** as a trigger to obtain the extremely high temperature required for fusion. In addition, an external **blanket** of uranium-238 undergoes fission from the neutrons produced by the fusion reaction. Thus, it is a fission-fusion-fission bomb.

gamma ray **Electromagnetic radiation** of very high energy. Gamma rays are one of the three radiations emitted by radioisotopes, and they can penetrate the human body. Shielding against them requires heavy metals, such as lead.

gene An entity that determines a hereditary property, such as eye color. A gene is now known to be a segment of the DNA molecule that determines the amino-acid sequence of a particular **protein**. Most of these proteins are the **enzymes**, which conduct most of the chemistry of the living cell.

genetic code The "language" used by **DNA** to determine **protein** structure. This language consists of three-letter "words," each letter corresponding to a sequence of bases (*A, T, G,* or *C*). Thus, the base sequence *CTG* in the DNA determines that the amino acid *aspartic acid* will be placed in the protein at that point.

geologic disposal The disposal of radioactive waste in rock formations deep in the ground.

germ cell The egg or sperm cell, and any precursor cell of the egg or sperm.

half-life (biological) The time required for half of a quantity of ingested atoms to be discharged from the body. Because elimination is a chemical process, all isotopes of the same atom have the same biological half-life. Cesium has a biological half-life of about 3 months.

half-life (physical) The time required for half of the radioactive atoms of a substance to decay (emit radiation). For example, cesium-137 has a physical half-life of 30 years. After 30 years, it will be 50% as radioactive; after 60 years, 25% as radioactive; after 90 years, 12.5% as radioactive, and so on. See Figure 6.

heat The degree of atomic or molecular motion.

heavy water Water in which some of the hydrogen is the heavy isotope **deuterium**.

hydrogen bomb (H-bomb) A **fusion bomb**.

ICBM *I*ntercontinental *b*allistic *m*issile. A **ballistic missile** of range up to 8000 miles, capable of going from one continent to another.

induced fission **Nuclear fission** produced by neutrons applied externally. Occurs in nuclear reactors and bombs. The term is used in contrast to **spontaneous fission**.

induced radioactivity Radioactivity induced in objects by bombardment with **subatomic particles** that are absorbed by the atomic nucleus. **Neutrons** are very effective in inducing radioactivity and account for the induced radioactivity at the site of a nuclear explosion.

inherently safe reactor A nuclear reactor so designed that it is virtually impossible for it to melt down, even if the operators are unusually inattentive. Several such designs exist, and some experimental inherently safe reactors have operated for years.

intermediate-range weapon A nuclear weapon with a range of 600 to 3000 miles.

ion Any atom with a net electric charge, caused by its having lost one or more electrons (becoming a positive ion), or having gained one or more electrons (becoming a negative ion). The chloride ion (Cl⁻) is a negative ion; the ferrous ion (Fe⁺⁺) is a positive ion.

ionizing radiation Any radiation that can remove an electron from an atom, thus converting it into an **ion**. Ionizing radiation has high energy and includes **gamma rays**, **X rays**, and **particle radiation**.

isolation Long-term safe disposal of **radwastes** in a manner that does *not* permit retrieval.

isotope An alternative form of an **element**, having the same **atomic number**, but different **atomic weight** (a different number of neutrons). *All* forms of an element, whether natural or not, radioactive or not, are isotopes of each other and have similar chemical properties.

kiloton Explosive force (not weight) equivalent to the explosive force of a kiloton (1000 tons) of TNT.

kinetic-energy weapon An antiballistic-missile weapon, such as an **electromagnetic rail gun**, that shoots material objects at high velocity.

lethal mutation A **mutation** that kills the cell or, in some cases, the organism in which it occurs.

linear hypothesis The hypothesis that the biological effects of radiation increase in direct proportion (linearly) to the **dose**.

Mutual Assured Destruction (MAD) The nuclear strategy of both sides having arsenals so great that the other side is assured of destruction if it attacks, resulting in a mutual deterrence of nuclear aggression.

mass A measure of the amount of substance in a body. At the surface of the Earth, mass is equal to weight.

megaton Explosive force (not weight) equivalent to the explosive force of a megaton (1 million tons) of TNT.

meltdown The melting of a **reactor core**, due to the fission reaction's going out of control. Meltdowns have occurred in several reactors, but the one at Chernobyl was the only one with severe biological consequences.

mill tailings The wastes of uranium ore after a mill has extracted most of the uranium. Though the level of radioactivity is low, the sheer volume of such wastes creates a problem.

MIRV *M*ultiple-headed *i*ndependently targetable *r*eentry *v*ehicle. An **ICBM** with more than one warhead, each of which can be aimed at a different target.

moderator A substance in the **reactor core** that slows down the fast neutrons from nuclear fission, without absorbing them, to produce the **thermal neutrons** required for further fission. Common moderators are graphite, water, and heavy water.

molecule A tiny structure made of two or more atoms held together by **chemical bonds**. A molecule of water (H_2O) contains two atoms of hydrogen and one of oxygen. Protein molecules contain thousands of atoms.

muon A kind of "heavy electron," with a mass roughly 200 times greater. Produced in the atmosphere by cosmic rays; also produced by high-energy particle accelerators.

mutation A permanent change in the hereditary material (**DNA**) of a living **cell**.

national technical means Methods of spying that use technical devices permitting observations from beyond enemy borders or from above the earth. Includes radar imaging and photography from space satellites.

natural background radiation Radiation that occurs normally in the environment. It includes radiation from such things as radon, cosmic rays, and radioisotopes inside the body.

NCRP National Council on Radiation Protection and Measurements (U.S.A.).

neutron A subatomic particle, with no electric charge (electrically neutral), that resides in the atomic **nucleus**. It has the same **mass** as the **proton**.

neutron activation analysis The detection and measurement of trace amounts of an unknown **element** in a substance by irradiating the substance with neutrons in order to induce artificial radioactivity.

neutron bomb   A **fusion bomb** without the outer uranium **blanket,** which therefore releases a much higher proportion of neutrons. Intended as an antipersonnel bomb.

neutron bombardment product   A usually radioactive **isotope** induced in the structures of a **reactor core** by intense bombardment by neutrons over the lifetime of the reactor. A typical such product is iron-55 (see Appendix C).

noble gas   A gas made of atoms (or the atoms themselves) that have a completed outer shell of electrons and that are therefore chemically unreactive. Examples are neon, xenon, and radon.

nonlethal mutation   A **mutation** that does not immediately kill the cell or organism in which it occurs. It may produce cancer or developmental abnormalities.

NRC   Nuclear Regulatory Commission, established in 1975. It performs the civilian licensing and oversight functions of the former Atomic Energy Commission (**AEC**).

nuclear bomb   A bomb whose explosion results from **nuclear fission** or from **nuclear fusion,** or both.

nuclear energy   The energy associated with the atomic nucleus, particularly as manifested by **radioactivity,** and by **nuclear fission** and **nuclear fusion.**

nuclear fission   The splitting of the nucleus of (usually) a very heavy atom, such as uranium or plutonium, into two daughter nuclei of roughly half the weight, with a concomitant great release of energy, partly in the form of energetic neutrons and gamma rays.

nuclear freeze   A proposed treaty that would stop all production, testing, and deployment of new nuclear weapons. Such an agreement would result in the gradual obsolescence of nuclear weapons.

nuclear fusion   The fusion of the nuclei of two very light atoms, such as **deuterium** and **tritium,** to form a heavier nucleus, such as that of helium. Occurs with a concomitant great release of energy, partly in the form of energetic neutrons and gamma rays. Ignition of the reaction requires extremely high temperatures.

nuclear medicine   The diagnosis or therapy of disease by the use of **radioisotopes.**

nuclear power   The harnessing of **nuclear energy** for the production of electric power.

nuclear reaction   Any reaction in which the composition of an atomic nucleus changes. Such reactions may be spontaneous, as in **radioactivity**, or they may be induced, as in reactors and bombs.

nuclear reactor   A machine in which **nuclear fission** occurs under controlled conditions, releasing heat that can be used for the production of electric power. Some reactors are designed primarily to produce specific isotopes.

nuclear winter   Cold weather resulting from blotting out of the sun by smoke and dust raised into the atmosphere by many nuclear explosions, as in a nuclear war. The effect could last for many months, or even years.

nucleus (atom)   The extremely tiny and massive core of an atom, which contains the **protons** and **neutrons**.

nucleus (cell)   A roughly spherical body, usually near the center of living cells. It contains the **chromosomes**, which possess the genetic material (**DNA**) of the cell.

overkill   The degree to which a nation's nuclear arsenal exceeds the amount needed to destroy the other nation.

particle radiation   Radiation consisting of **subatomic particles**, such as **protons, neutrons, electrons,** and **alpha particles**.

periodic table   A table that lists all the **atoms** (or **elements**) that exist, in the order of their **atomic numbers**. Atoms in the same vertical column of the periodic table have similar (but not identical) chemical properties.

phased-array radar   An early-warning radar that can sweep across the sky electronically, rather than by mechanical rotation of the antenna, and is therefore extremely fast.

photon   The unit (or "quantum") of **electromagnetic radiation**. Photons behave like particles, but they also behave like waves, a paradox not easy to understand.

plutonium   A fissionable **transuranic atom** of **atomic number** 94, two atoms beyond uranium in the **periodic table**. It does not occur in nature. Plutonium is a by-product of **nuclear fission** in uranium that contains uranium-238. Can be used in **nuclear bombs** or as **nuclear reactor** fuel.

plutonium production reactor   A military **nuclear reactor** designed to produce weapons-grade (high purity) plutonium.

postboost phase The period in the trajectory of a ballistic missile after the **boost phase**.

pressure The *force per unit area* exerted on a container by the moving molecules inside.

pressurized-water reactor (PWR) A **nuclear reactor** in which the cooling water is maintained at such a high pressure (over 100 times atmospheric pressure) that it cannot boil. The very hot water from the reactor is passed through a heat exchanger, where it heats a *secondary* water system that is permitted to boil and thus produce steam to power the turbines.

probability A mathematical expression of the chance that something will happen, measured on a scale from 0 to 1. Impossible events have a probability of zero, while absolutely certain events have a probability of 1.

proliferation The proliferation of nuclear-energy knowhow and/or equipment from the nuclear-capable nations to non-nuclear nations. The concern is that more nations would become capable of making nuclear weapons.

protein A linear polymer of amino acids that acquires either a spherical form, as in **enzymes** and antibodies, or a fibrous form, as in keratin and collagen.

proton A subatomic particle with a unit positive electric charge that resides in the **nucleus** of the atom. It has the same **mass** as the **neutron**.

radiation therapy Treating a disease with radiation, such as **X rays** or radiation from **radioisotopes**.

radioactive tracer A **radioisotope** fed to a biological system so that it will go to its normal site in the system, which site can then be identified by the emitted radiation.

radioactivity The spontaneous emission of radiation (**alpha rays, beta rays,** or **gamma rays**) by an unstable atomic nucleus. The instability results from an excess of **neutrons** in the nucleus.

radioisotope An **isotope** that is radioactive (exhibits radioactivity).

radwaste Radioactive waste.

reactor core The heat-producing part of a **nuclear reactor**. It contains the **fuel rods, control rods,** and **moderator**. It is typically surrounded by a *containment vessel* of thick steel.

recessive mutation A **mutation** that is observed in the organism only if it occurs in the same gene on both members of a pair of chromosomes. Because such a double occurrence is rare, a recessive mutation is usually not expressed. However, it may show up in subsequent generations, by a mating with another organism that has that same mutation.

reentry phase The period in the trajectory of a ballistic missile when it passes through the atmosphere to its target. This phase lasts about 30 to 100 seconds.

rem *R*oentgen-*e*quivalent-*m*an. A measure of radiation **dose** that enables one to compare the biological effects of different kinds of radiation, such as X rays and neutrons. (See Appendix B.)

reprocessing The treatment of **spent fuel** from a nuclear reactor to extract the fissionable uranium and plutonium that can be used again.

rocket A propulsion device that acts on the principle of reaction to the explosive force of a fuel. Rockets work better in space than in the atmosphere, whereas a jet engine, requiring air, cannot function in space.

silo An underground storage place from which an **ICBM** can be launched.

SLBM Submarine-launched ballistic missile.

somatic cell Any cell in the body other than the **germ cells**.

spent fuel The used-up fuel from a nuclear reactor. Spent fuel no longer functions efficiently because of depletion of its fissionable fuel (typically uranium-235) as well as neutron absorption by accumulated fission products. In a nuclear power plant, such fuel is about 3 years old.

spontaneous fission The natural (spontaneous) **nuclear fission** of a very heavy atom. Of natural atoms, only uranium and thorium show spontaneous fission.

Star Wars A somewhat derogatory term applied to the **Strategic Defense Initiative**.

Strategic Defense Initiative (SDI) A strategy enunciated by President Reagan in 1983 in which an extensive and complex system of mostly space-based weapons would be deployed to shoot down incoming nuclear missiles.

strategic nuclear weapon A nuclear weapon of long range, capable of attacking the enemy homeland.

subatomic particle A particle that is a constituent of an atom, such as a **proton** or an **electron**.

tactical nuclear weapon A weapon of shorter range, for use on the battlefield as well as in attacking rear formations and installations.

telemetry The radio broadcasting of test data from an experimental rocket or ballistic missile. The encoding of such data, so that it cannot be interpreted by an outside nation, is a bone of contention between the U.S. and the USSR.

thermal neutron A **neutron** with energy comparable to that of molecules at room temperature. Such "slow" neutrons are required for the efficient **nuclear fission** of uranium-235.

thermal pulse The wave of intense heat that spreads out from a nuclear explosion. Consisting of visible light and infrared radiation, it travels at the speed of light and can last as long as 10 seconds for a 1-**megaton** explosion.

thermonuclear reaction **Nuclear fusion** reaction. So called because of the extremely high temperatures required for ignition.

thermonuclear weapon A weapon carrying a thermonuclear, or **fusion bomb**.

transuranic atom An atom beyond (of higher **atomic number** than) uranium in the **periodic table** (Appendixes A and C). These atoms do not occur in nature; they are all man-made. Example: **plutonium**.

tritium A heavy isotope of hydrogen, with one **proton** and *two* **neutrons** in its nucleus. It is radioactive (half-life 12 years). Small amounts are constantly produced by cosmic radiation acting on water vapor high in the atmosphere.

uranic An atom associated with uranium, including uranium, thorium, and their **decay products**, such as radium, radon, or lead (see Appendix C).

warhead The explosive part of a weapon.

waste (high-level) **Radwaste** that poses a long-term health hazard, having a high **activity** and a long **half-life**. Includes **spent fuel** and wastes from **reprocessing** plants.

waste (low-level) **Radwaste** that does not pose a long-term health hazard, having either a low **activity** or a high activity with a short **half-life**. Includes virtually all medical, industrial, and research wastes, as well as **mill tailings**.

X ray **Electromagnetic radiation** of energy between ultraviolet light and **gamma rays**. It is **ionizing radiation**.

⚛ Index

All entries followed by a "G" can be found in the Glossary.